A GUIDE TO THE FERNS
OF GREY AND BRUCE COUNTIES, ONTARIO

The Bruce-Grey Plant Committee
(Owen Sound Field Naturalists)

This book includes most of the ferns found in Ontario

Reprinted 2002 with revisions

Stan Brown Printers Limited
Owen Sound, Ontario

Canadian Cataloguing in Publication Data

Main entry under title:

A guide to the ferns of Grey and Bruce Counties, Ontario

Includes bibliographical references and index.
ISBN 0-9680279-2-X

1. Ferns – Ontario – Grey (County) – Identification. 2. Ferns – Ontario –
Bruce (County) – Identification. I. Owen Sound Field Naturalists. Bruce-
Grey Plant Committee.

QK525.7.O5G84 1999   587'.3'0971318   C99-930742-8

Cover Photo: Nelson Maher. Cinnamon Fern in Holland Township, Road 1, Cinnamon Bog.

Other publications by the Bruce-Grey plant committee available from the Owen Sound Field Naturalists:
A Checklist of Vascular Plants for Bruce and Grey Counties, Ontario, 2nd Edition 1997
The Orchids of Bruce Grey and Counties, Ontario, 1997
The Asters, Goldenrods and Fleabanes of Grey and Bruce Counties, Ontario, 2000
Rare & Endangered Species of Grey and Bruce Counties, Ontario, 2001

All books available from:
     The Bruce-Grey Plant Committee
     c/o The Owen Sound Field Naturalists
     Box 401, Owen Sound, Ontario N4K 5P7

Fern Guides may also be obtained from:
     The Toronto Field Naturalists
     2 Carlton Street, No. 1519
     Toronto, Ontario M5B 1J3
     Telephone: 416-593-2656
     Office open Fridays 9:00 a.m. to 12:00 noon

TABLE OF CONTENTS

PREFACE

This book has been produced in a cooperative effort by the members of the Bruce-Grey Plant Committee, all of whom have brought their different skills and knowledge to this project. They are acknowledged below. They have been assisted by many other individuals and organizations whose help is gratefully recognized here. No financial assistance has been received from any governmental agency or corporation. Since Grey and Bruce Counties are a wonderful area for ferns, the purpose of this book is to provide, for amateurs and professionals alike, a reference which focuses on the identification of our local ferns. The intent is to increase overall awareness of the botanical uniqueness of this area and the extreme importance of preserving suitable habitat for these fascinating plants while we still have the opportunity to do so. This book will be found useful in many parts of the province, as approximately 67% of the fern species present in Ontario are described here.

Many people are interested in growing ferns. This is best done by collecting spores, germinating them, and following their development (see page 110). Bear in mind that most ferns have very specific habitat requirements, and to grow them successfully one must recreate the right conditions. On no account should ferns ever be dug up and removed from wild areas unless it is certain that they are about to be destroyed by development or road construction.

ACKNOWLEDGEMENTS

* Members of the Bruce-Grey Plant Committee

* **Nels Maher** for being the source of inspiration for this book. An enthusiastic naturalist in the Owen Sound area from his earliest youth, Nels and his brother-in-law John Weir became interested in naturalized fern gardens about twenty years ago, and Nels is now the recognized expert on ferns in this area. He guides fern field trips for the Federation of Ontario Naturalists and many other nature groups, and with the aid of his wife Jean, has promoted much interest in this fascinating group of plants. He has been a fountain of information on the ferns of Grey and Bruce Counties, including their distribution and relative abundance. He was responsible for the major part of the writing, proof reading and editing of this book. His collection of silhouettes, produced by him from specimens that he collected, and his photographs, have provided the illustrations. His long experience as a printer has been applied to the layout and production of this book.

Donald Britton of the University of Guelph for expert advice and information on species distribution, for checking identification, and for thoroughly reviewing the manuscript during its preparation and making helpful suggestions.

* **Mac Campbell** for sharing his records of fern locations and assisting in field work.

Lenore Carrick for diagrams.

* **Joan Crowe**, project coordinator, for assisting in fieldwork, for writing some sections and for typing, proof reading and editing the text as a whole.

Walter Crowe for computer assistance, photographing specimens of rare species at the R.O.M. and for assisting with proof reading.

Tim Dickinson and **Deborah Metsger** of the Toronto Herbarium (TRT) at the **Royal Ontario Museum** for permission to search the fern collections and for assistance in photographing and photocopying specimens.

Marg Gaviller for help with proof reading.

* **Bob Gray**, Owen Sound Area Ecologist for the Ministry of Natural Resources, for procuring and supervising the student assistant, for writing the introduction to the area and the section on rare ferns, for proof reading and for making available fern records from M.N.R. reports on Areas of Natural and Scientific Interest (ANSI) and other documents.

* **Ellen and Orris Hull** for assistance with administration, for many helpful suggestions, for proof reading and assistance with editing.

* **Joe Johnson** for expert advice on descriptions of ferns in this area, their habitats, distribution, local abundance, and rarity; also for proof reading and editing. Many of the MNR studies and the ANSI reports consulted were his work.

Mac Kirk for encouragement and information on fern distribution.

Phil Kor, Conservation Geologist, MNR Lands and Natural Heritage, for writing the section on the Geology of Grey and Bruce Counties.

Jean Maher for assistance with fieldwork and proof reading.

* **Kathy Parker** for collating the records of the parks and reserves managed by McGregor Point Provincial Park and making them available to us, for testing the identification section, and for proof reading.

Martin Parker for records from Bruce County and helpful suggestions with respect to the survey of southern townships conducted in 1998.

Ian Sinclair summer student 1998 for willing assistance with the fern survey of southern townships and surveying some sites unsupervised; for many helpful suggestions as to possible locations; for keeping records and indexing the manuscript.

Dave Sinclair for accompanying and assisting Ian on the survey.

Stan Brown Printers staff for their expert assistance and willing cooperation in completing this project.

The Owen Sound Field Naturalists and the Saugeen Field Naturalists for their continuing support and encouragement.

Introduction To Grey and Bruce Counties

Location and Area

Grey and Bruce Counties are situated at the northern limits of Southwestern Ontario. Their latitude ranges from 43° 54' 30" N (just southeast of Whitechurch, near Lucknow) to 45° 19' 40" at Gig Point on Cove Island off the northern tip of the Bruce Peninsula. Roughly triangular in shape, the area is bordered by Lake Huron on the west and Georgian Bay on the northeast, providing some 750 kilometres of Great Lakes shoreline. Huron and Wellington Counties are to the south and Dufferin and Simcoe Counties to the east. The land areas of Grey and Bruce Counties are roughly one million acres each.

General Description

The Grey-Bruce area is known for its natural beauty and clean environment. It includes the Niagara Escarpment, the Bruce Trail, Sauble Beach, a world class cold-water fishery in Georgian Bay and Lake Huron, wild orchids and the near wilderness area of the Bruce Peninsula which is still home to the Black Bear and the Eastern Massasauga Rattlesnake.

Bruce County is characterized by a transition from pasture and flat cropland in the south to the rugged, primarily forested lands of the Bruce Peninsula. Grey County is a patchwork of woodlands, wetlands and agricultural lands. It has a varied topography, with numerous cold water streams and inland lakes. Northern and eastern Grey County and northern Bruce County are dominated by the Niagara Escarpment.

Geology

Southern Ontario is blanketed with sedimentary and carbonate rocks of Paleozoic age. If one drills a deep enough hole through these rocks, one will eventually reach the rocks of the ancient Canadian Shield. These basement rocks, some 2.5 billion years old, are the same rocks that are exposed in the Parry Sound-Muskoka region across Georgian Bay, and that also form the platform for much of Northern Ontario.

The relatively young Paleozoic rocks of southern Ontario are between 360 (Devonian) to 450 (Ordovician) million years old. They were deposited in a complex warm-water, tropical sea occupying the landlocked environment of a large saucer-shaped depression, the Michigan Basin, centred on the State of Michigan. The rocks exposed in Grey and Bruce Counties were formed at the northeast edge of this basin, where the water levels were quite shallow. The rocks were originally muds, silts and sands, and, particularly, coral reefs. Later sediments include fragmented coral. The Paleozoic sediments were deposited originally as flat-lying, horizontal strata that have been largely preserved to the present day, despite slight alterations caused by compression over time, and by uplift of the underlying Precambrian basement. This latter event formed the

Algonquin Arch, and lifted the rocks of the region to form a dome on which one of the highest points in southern Ontario is found. The highest point along the Niagara Escarpment occurs in the Grey-Bruce area, at the Blue Mountains, noted for its ski hills. Here, also, the Paleozoic sequence is fully exposed — down to its oldest layers. The rocks of southern Ontario are largely covered by loose glacial sediment laid down by the glaciers and glacial rivers of the last Ice Age. The notable exception is the Niagara Escarpment which, in its vertical cliffs, exposes one of the best continuous sections through Silurian (430-400 million years ago) strata anywhere in the world.

Numerous other areas of exposed rock outcrop in the Grey-Bruce area, away from the Escarpment. This is due in part to powerful erosion by glacial ice and water which removed much of the existing sediments from the area some 12 000 years ago. Examples of this are to be seen in the deeply cut beds of modern rivers, such as along the Rocky Saugeen River in Glenelg and Bentinck Townships and the upper Teeswater River in Culross Township. Another well known exposure is at the Formosa North Road Cut near the community of Formosa.

Most of the exposed bedrock in Grey and Bruce Counties is dolostone. Some of the rock pavements in the upper Bruce Peninsula are excellent examples of alvars. Water and chemical erosion of the exposed bedrock near the Escarpment in Grey-Bruce has resulted in the development of karst topography. Karst is the pitting, smoothed channels, blocks, passageways and caves found in some of the dolostone formations in the region. Some of the best karst topography in Canada occurs at the north end of the Bruce Peninsula, at Cabot Head.

Physiography

The Grey-Bruce area has been covered by continental glaciers many times during the last two million years or so. The last ice sheets retreated from the area a mere 10 000 to 12 000 years ago. Powerful erosion by glacial ice and meltwater, and the subsequent inundation by high water levels in the Georgian Bay and Lake Huron basins had great effect in shaping the landscape we see today. The most obvious effects of glacial events in Grey-Bruce are the deep narrow valleys which were cut through the Niagara Escarpment by lobes of ice and meltwater floods beneath the ice. The heads of these valleys form natural harbours, and are now occupied by communities such as Owen Sound and Wiarton. Where the valleys were filled in with sediment, such as in the Beaver Valley, they are now occupied by much smaller modern streams known to geographers as "misfit" streams which flow through the old valleys down to modern lakes.

Much of the Bruce Peninsula is typified by shallow soil to no soils. This is the result of the removal of existing material by powerful floods from beneath the last glaciers, as well as erosion by the ice itself. Deeper soils remain in southern Grey and Bruce Counties. In Arran and Sydenham Townships, the soils have been drumlinized by the effects of the passing of glacial ice and meltwater. Parallel ridges of jumbled sediment, trending east-west across the area, mark recessional moraines where the retreat of the glaciers was halted for a time. Eskers, long ridges of sand and gravel, tending to be parallel to the direction of ice movement — north to south here — represent sediments deposited in rivers beneath the glaciers. Adjacent to the ice front and away from it, large rivers of meltwater carried off the sediments and carved the wide shallow channels we find today.

The Lake Huron shoreline is well known for its numerous calcareous fens and the rare species associated with them. Sandy beaches and associated dunes also occur along the Lake Huron shoreline from Point Clark to the Howdenvale area. When lake levels were much higher following the retreat of the glaciers, the conditions for the formation of similar features existed. Old beaches, spits, sandbars, cliffs, and stacks ("flowerpots") long since abandoned by the original shoreline on which they formed, occur parallel to, and inland from, the present shoreline. At the very top of the cliffs at Cabot Head, cliffs and flowerpots which formed on the shores of tiny islands in glacial Lake Algonquin some 12 000 years ago are delicately preserved some 66 metres above the present level of Georgian Bay.

Much of Grey and Bruce Counties is imperfectly drained leaving many wetlands including the largest remaining wetland in southern Ontario, Greenock Swamp.

Climate

Lake Huron and Georgian Bay, with their prevailing westerly winds and on-shore breezes, have an appreciable moderating effect on the climate of Grey and Bruce Counties. The cold water keeps air temperatures cool in the spring, normally retarding flowering until the danger of frost is past. In the autumn, when the water is still relatively warm from the high air temperatures of summer, the lakes tend to keep the adjacent areas warmer, thus extending the frost-free period. The lakes also provide a more than adequate water source for precipitation throughout the year. This moderating effect is one of the key reasons why the Meaford-Thornbury area of southern Georgian Bay, with an average of 140 frost-free days, is the top apple producing region of the province. Inland, away from the moderating effects of the water and with a considerable increase in altitude, the frost-free days are fewer. Durham, for example, on average has only 115 days without frost. Thus, the climate throughout Grey and Bruce Counties shows considerable local variation.

Human History and Land Use

This area was one of the last parts of southern Ontario to be settled by Europeans. With the exception of the early French missionaries, it was essentially unexplored by white people until 1815 when Captain William Fitzwilliam Owen undertook a hydrogeographic survey of Georgian Bay and Lake Huron. However, little was known of the land which was controlled by the Ojibway. In 1818, they turned over the easternmost part of Grey County to the Crown, but the major portion of the two counties (the Queen's Bush) was not surrendered by the Ojibway until 1836. European settlers began arriving by about 1840, and the area was soon dotted with communities and small farms. The Bruce Peninsula, except for some relatively small reserves, was given up by the aboriginal people in several stages between 1854 and 1885. In 1855, a railway line came to Collingwood from Toronto, and by 1873, Owen Sound also had a rail link to the south. These lines led to greatly increased shipping on Georgian Bay which, in turn, helped to make the area more accessible.

The main occupation of the early settlers consisted of clearing and seeding new land. The families produced almost everything for their own needs. Wheat, hay and potatoes were the main crops. Food for the family and livestock were produced during the summer. In winter, lumbering for cash was the main occupation. By the 1880's most farming was mixed, but many of the farmers were preparing to convert to raising livestock exclusively. Today, the major products are apples, beef cattle, sheep, mixed grain, hay and dairy.

The Bruce Peninsula was thoroughly logged in the late 19th and early 20th centuries and considerable portions then burned. Much of the land was found to be unsuitable for agriculture. The forest cover of Grey and Bruce Counties is still approximately 40% overall, varying approximately from 7% in Huron Township in southern Bruce to 85% in St. Edmunds Township in the northern Bruce Peninsula.

Vegetation

The natural vegetation of Grey and Bruce counties was primarily forest. There would have been some natural openings such as the sand dune areas adjacent to Lake Huron, alvars and limestone pavement areas associated with the northern Bruce Peninsula, floodplain areas of the larger rivers such as the Saugeen, and the numerous unforested wetlands that are found along the shorelines as well as inland.

Drainage and soil type is the key to the kinds of forest that originally occurred here. Most parts of Grey and Bruce Counties were (and still are) dominated by Sugar Maple forests. White Pine was very common throughout the area and Red Oak occurred frequently on drier sites. Wet areas supported mixed forests of White Elm, Red Ash, Aspen, and White Cedar. Eastern North America's most ancient White Cedar forest still clings to the inhospitable cliffs of the Niagara Escarpment along its course through Grey and Bruce Counties. Multiple uses from forestry agriculture to downhill skiing have fragmented the original forest and, in many cases, changed the species composition. Most of the present day forest is second growth. Nevertheless, this area has more forest than most other parts of southern Ontario.

Tobermory

Bruce
Peninsula

Georgian Bay

Wiarton

Lake Huron

Meaford

Bruce County

Grey County

Owen Sound

Southampton

● Markdale

●
Walkerton

Why are there so many species of ferns in Grey and Bruce Counties?

Within Grey and Bruce counties 50 species of ferns are documented as currently occurring. This is 3½% of our total flora whereas the 75 ferns recorded for the whole of Ontario comprise only about 2⅓% of the total flora. Only two fewer ferns are found here than in the entire British Isles. For such a comparatively small area, Grey-Bruce has a remarkably high percentage of the list of ferns for the whole province. This is due to a combination of factors. First, even though it takes about 2½ hours to drive from the southern boundary of Bruce County to the northern tip of the Bruce Peninsula, there is a considerable change in latitude which is reflected in the climate and the resultant natural vegetation. This is accentuated by the modifying effect of Lake Huron and Georgian Bay, especially in the Bruce Peninsula.

The Grey-Bruce area is located entirely within the Great Lakes–St. Lawrence Forest Region which is a transition zone between the Deciduous or Carolinian Forest to the south and the conifer dominated Boreal Forest to the north. The mixing of northern and southern species is, for the most part, quite subtle but it becomes most obvious in St. Edmunds and Lindsay Townships at the northern end of the Bruce Peninsula where a major component of the forest is Jack Pine — nearly at its most southerly limit in Ontario. Examples of more southerly species reaching the northern edge of their range in Grey-Bruce are Hackberry, Blue Beech and Bitternut Hickory. Similarly, some of the plants of the forest floor, including ferns, are at the limit of their range. Northern species of fern at their southern limit in Ontario include Robert's Fern and Laurentian Bladder Fern. Southern species at their northern limit include Narrow Leaved Glade Fern, Goldie's Fern and Clinton's Wood Fern.

The Niagara Escarpment is the dominating feature in northern Grey County and on the Peninsula. This landform creates a tremendous variety of microhabitats. Weathering of the exposed dolostone creates a continuous supply of calcium. A significant proportion of our ferns are calciphiles which will flourish only in highly alkaline situations. Examples of these are Bulblet Fern, Green Spleenwort and Walking Fern. Two calciphiles which are extremely rare in North America, Wall Rue and Hart's Tongue Fern, occur along the escarpment. The relatively high percentage of ferns in the total flora of Grey and Bruce Counties is primarily due to the great amount of exposed, or nearly exposed, dolostone. In combination with a heavy protective snowfall this gives the area an advantage over other parts of Ontario. Many ferns are very exacting in their habitat requirements, and the great variety of landforms within this relatively small geographic area provides a wide diversity of conditions. Apart from the general modifying effect of Lake Huron and Georgian Bay, the predominantly westerly winds and on-shore breezes bring cool moist air to the shoreline areas which have a direct impact on microclimate in the various habitats. The variety of exposure created by the wetlands, forests and rocky uplands affects light levels and moisture levels producing variations in microhabitat which are critical for many fern species.

Past and present land use of the area is also an important reason why so many ferns occur. There is still a high percentage of forest cover in the two counties. Although the land was divided up for farming, much of it was either too wet or too rocky. Many marginal areas that were cleared for agriculture have been abandoned and are reverting to forest. Thus, much habitat suitable for ferns has been preserved or is regenerating.

The amount of effort exerted in discovering and documenting ferns in this area since the land was ceded by the Ojibway is undoubtedly a factor in the high number of ferns recorded. The Bruce Peninsula has, for over 100 years, been renowned as a botanical treasure house, and has been thoroughly explored by numerous, often famous, botanists including the Canadian John Macoun. Most other parts of Grey and Bruce Counties have not been so well studied, and more work is needed to record the distribution of ferns, even on a township unit basis. Much of the remnant forest in the south is on private land. Any information about fern discoveries not recorded in the township unit chart should be passed on to the Owen Sound office of the Ontario Ministry of Natural Resources, 1450 Seventh Avenue East, Owen Sound, ON N4K 2Z1. (See reporting form at the back of this book).

The Grey-Bruce area is very fortunate in having so much public land which is owned and managed by various organizations including Parks Canada, Ontario Parks, Ministry of Natural Resources (Crown lands), Conservation Authorities, The Federation of Ontario Naturalists, The Bruce Trail Association, The Ontario Heritage Foundation, Canada Coast Guard, as well as municipal lands such as County Forests and township and city parks. In addition, access is provided to some private lands by participating landowners along the Bruce Trail. We are indeed favoured by having so many places that we can explore and discover ferns in this unique area. It is up to all of us to respect these lands and do everything we can to preserve them in their natural state. It is particularly important to respect private property and to obtain permission of the owners if you wish to visit their land.

INTRODUCING FERNS

Ferns are a very ancient group of plants, but they have maintained their genetic flexibility to this day; so they are still capable of throwing up unexpected variations and hybrids to surprise the unwary botanist. Darwin said that the fossil record was like "a history of the world imperfectly kept, and written in a changing dialect; of this history we possess the last volume alone .... of this volume, only here and there a short chapter... and of each page, only here and there a few lines." By chance, the ferns are, in this respect, one of the better documented groups. It is possible to trace their development from some of the first land plants in the Silurian period (430 to 400 million years ago). These were erect, leafless and rootless spore bearing plants. In common with mosses and other bryophytes, these plants had in turn arisen from aquatic algae of some kind, as conditions changed in such a way that they were able to invade and colonize the land, a process which took many millions of years. All these groups still share many features in their life cycles and biochemistry. By the Devonian period (400 to 360 million years ago), an immense variety of ferns were to be found on earth and this continued into the Carboniferous periods when the coal seams were laid down. The moist, humid conditions of rapid deposition preserved ferns so perfectly in layers of coal, that it is just as if someone had taken the trouble to press them. Of course, at this time ferns were the dominant form of plant life on land. Then came the ice ages of the Permian (286 to 245 million years ago) with massive extinctions, but as usual, some species survived both this and the later Pleistocene glaciations, and have continued with little change in appearance down to the present day. The number of fern species in the world today has been estimated at 9000. They range in variety from the tree ferns, mainly tropical, through the vast spreads of bracken found on the denuded hills of temperate Europe, the carpets of Ostrich Fern in our damp bottom lands, tiny filmy ferns on shaded, moist rocks and tree trunks in more southern climes, to truly aquatic species like Water Clover.

Most of us were introduced to ferns as plants that do not flower but reproduce by spores. While it is true that the plants we recognize as ferns are the **sporophyte** generation which produces the spores, there is a second type of plant which develops from the germinating spores. This is the **gametophyte or prothallus**. It is a tiny insignificant sheet of green plant tissue that you would be very lucky to find. Nevertheless, they are not difficult to grow if you collect some fern spores and keep them in a moist, dark place (see p. 110). The cell division that produces the spores is known as **Reduction Division** because after this division each spore will contain only half the chromosomes (n) of the parent plant. The **haploid** gametophyte, derived from the spore, produces the egg cells and the sperm cells. The latter must swim through a film of water for fertilization to take place. The **diploid** sporophyte — with two full sets (2n) of chromosomes — then develops from the fertilized egg in the tissue of the gametophyte which eventually disappears, leaving the sporophyte fully independent.

LIFE CYCLE OF A FERN

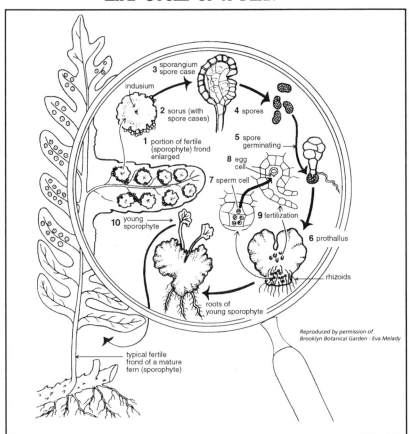

FIGURE 1

Reproduced by permission of
Brooklyn Botanical Garden - Eva Melady

Fern sporophytes are well adapted to life on land. The sporophytes of some species grow in remarkably arid places, and the dispersal of spores is usually by wind. The gametophyte, however, is a very simple structure which cannot withstand excessive drying, and the requirement of water for fertilization is a limiting factor for ferns. Consequently, ferns are not so independent of water as the flowering plants with their more advanced methods of pollination and seed development. Ferns do have a fully developed root system and vascular tissue in the stem and leaves. This allows conduction of water to all parts of the plants but also serves as a support system, enabling ferns to grow much taller than non-vascular plants such as mosses. Many ferns have a tough rhizome, or underground stem, which anyone who tries to move an Ostrich Fern will discover. This enables the plant to survive grazing animals, drought and winter cold. Many die off above ground in winter and regenerate new leaves from the rhizome in the spring, but some of our species have evergreen leaves that persist under the snow. In warmer countries the tough stems of some species rise above the ground forming the "trunk" of the tree ferns. Nevertheless, ferns do not produce true "wood" as do our deciduous and coniferous trees.

PARTS OF A FERN

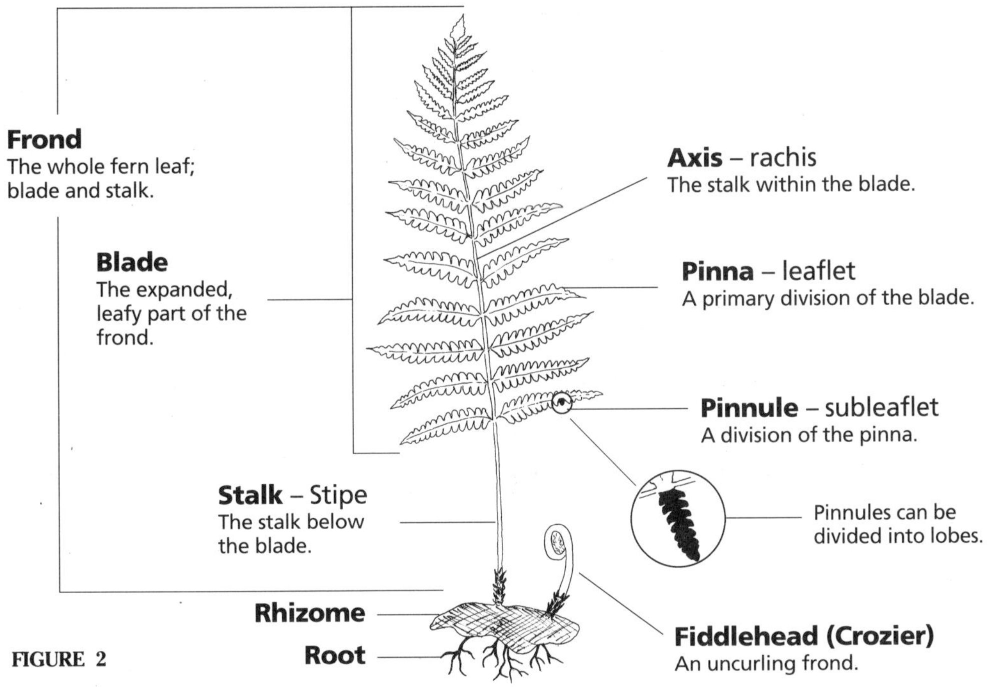

Frond
The whole fern leaf;
blade and stalk.

Axis – rachis
The stalk within the blade.

Blade
The expanded,
leafy part of the
frond.

Pinna – leaflet
A primary division of the blade.

Pinnule – subleaflet
A division of the pinna.

Pinnules can be
divided into lobes.

Stalk – Stipe
The stalk below
the blade.

Rhizome

Root

Fiddlehead (Crozier)
An uncurling frond.

FIGURE 2

The structure of ferns is somewhat different from that of flowering plants. The leaves are usually referred to as **fronds**. They are supported by a **stalk** or stipe which may be scaly. It continues as the **axis** or rachis, this is the central rib of the **blade**. The blade may be simple as in the Hart's Tongue Fern. In some, such as the Christmas Fern, it is **divided once** into **pinnae** or leaflets. The edges of the pinnae may be wavy as in the Sensitive Fern but if they are clearly divided into **pinnules** or subleaflets, the leaf is considered to be **twice divided** e.g. the Silvery Glade Fern. The pinnules, in their turn, may be lobed again (**thrice divided**) e.g. Intermediate Fern. In some ferns, such as Bracken and Oak Fern, the axis appears to divide into three, producing a compound or **ternate** frond because the lower pair of pinnae are very large compared to the rest of the pinnae. These features enable us to split ferns, artificially, into manageable groups for purposes of identification, although they do not represent the true relationships between different species.

The terms in bold print will be the ones used consistently in the descriptions in this book. A glossary defining all terms will be found at the back of the book on page 111.

The dustlike spores are produced by the sporophytes in tiny structures called **sporangia**. These are designed to open in dry weather and close under wet conditions, thus facilitating spore dispersal on air currents. They usually form in clusters called **sori** (singular — sorus) on the underside of the frond or in the rolled under edge of the pinnae. The sori may be protected by a sheet of tissue called an **indusium**. Where they are grouped in spots on the underside of the fronds, they are often referred to as "fruit dots." This is a very inaccurate term as there is no similarity between a fruit, which is a ripened ovary containing seeds, and a group of sporangia containing spores. A somewhat more accurate term is "spore dot." Fern spores may travel great distances but they are not very long lived. Whether they germinate or not depends on their landing in habitat suitable for that particular fern and with the right climatic conditions.

Another characteristic of most ferns which distinguishes them from flowering plants is the curious way in which the young leaf is rolled up so that the stalk arches over and protects the blade. At this stage, the stalk is usually very hairy or scaly, providing further protection to the developing leaf. This circular type of development is known as **circinnate vernation**. The resemblance of the coiled leaf to the scroll of a violin has led to the term **fiddleheads**, being used in reference to the developing leaves, especially as they emerge in the spring. The word **crozier** is also applied, having arisen from the resemblance to a bishop's crook. This is probably more applicable to some of the smaller, more wiry ferns where the stalk is longer and the blade forms a very small coil. To avoid confusion we are using the word fiddlehead in all the descriptions.

Unlike smaller plants such as mosses and algae, many ferns are well recognized by ordinary people and so have been given English, or common, names. These may vary from place to place, or in some cases, the same name may be used for two different species. For scientific purposes, every plant has a **binomial** Latin name giving the genus and species e.g. the Cinnamon Fern is *Osmunda cinnamomea* L. The letter or syllables after the species name designates the author who first named the plant — in this case, the eighteenth century Swedish botanist Linnaeus (Carl von Linné) who named thousands of plants from all over the world. Sometimes the name is modified in some way at a later date. Then a second author name may appear after the first e.g. the Purple Stemmed Cliffbrake is *Pellaea atropurpurea* (L.) Link, the latter being the accepted abbreviation for the German botanist Joseph Link (1767-1851). This is an easily recognised species, and the name was settled a long time ago. There has been considerable controversy about a number of our groups of closely related ferns as to whether these ferns are separate species or simply varieties of species in the group.

Many of our species have arisen, by crossing between two (diploid) species and subsequent doubling of the chromosomes. In some cases, at least one of the parent species still occurs in its original form and there are similarities between the two. Many once-widespread species distributions were dissected by the Pleistocene glaciations, and it was a matter of chance where a particular species survived and whether it was able to recolonize an area after the ice retreated. The pattern of extinctions in Europe, where the Alps form a barrier, differed from that in North America where the north-south orientation of the mountains provided an escape route. When botanists started naming species in a systematic fashion, it was not uncommon for the same species to be given one name in Europe and another in North America. There is now an agreed International Code of Plant Nomenclature which dates from 1956, but it has taken a long time to sort out all the problems created by earlier lack of communication. Recent chromosome and computer aided studies have done much to clarify the situation. Amateurs do, however, find it very confusing when every book they pick up seems to have different names for the same fern, not to mention a different system of names for the parts of the plant! The English names given in this book are, generally, those in current use in Grey and Bruce Counties. The Latin names correspond to those in the Flora of North America Vol. II 1993 but common synonyms, likely to be found in older books, are also included. Hybrids will be mentioned, where appropriate, in the text. Their Latin names contain an "**x**" signifying that two species have crossed.

IDENTIFYING FERNS

The characteristics used to identify ferns are often not as obvious as those found in flowering plants. Colour, for instance, is generally in shades of green and brown, but it should be observed carefully because it often helps to distinguish closely related species. In ferns, the frond shape is extremely important. The shape of the outline should be noted, and also, whether the blade is subdivided once, twice or three times. The margins of the subdivisions may be toothed. The stalk, too, may give important clues from its colour or whether it is smooth, hairy, scaly or glandular. The axis should be similarly observed. Lastly, the spore producing structures are very characteristic. In order to observe these clearly, a hand lens (magnifier) giving 10 x magnification is essential. It should be strung around your neck on a cord as it is extremely easy to lose in the woods, especially if you drop it in a dense stand of ferns or in a creviced, rocky area! Ferns show a lot of variability even in one species, and beginners may find this confusing. When you have only one specimen in front of you and nothing to compare it with, you will often have difficulty being sure which species it is. However, experience helps solve a lot of these problems and there are only 50 species to get to know, of which at least half are easily distinguished.

A field note book is another useful tool. Noting all the features you see in a systematic way will help you sort them out mentally. Sometimes, it may be necessary to take a frond or a piece of a frond; as long as there is a good stand of plants this will do no harm. However, if there are only ten or fewer plants of that species in the area, they should not be touched. If you do collect a frond, it should be pressed immediately and firmly in newspaper and dried quickly. Many ferns wilt rapidly when picked. Photographs may be useful but only if you make careful notes about shape and size. Whether taking a photograph or collecting a specimen, the precise location and detailed habitat information should be recorded. Habitat information can help with identification. For example, some species grow only on rock, others only in wet areas. In addition, if by chance you have found a species not previously recorded in the area, detailed background information is invaluable. Exercising your powers of observation in this way will give you a heightened sense of perception when you are outdoors. You will start to see things you never knew existed!

Aids to Identifying Ferns

The following diagrams will clarify what you need to look for. After that, the list of species has been broken down into smaller groups under various headings to assist you in getting started on identification. The silhouettes with the species descriptions and the photographic section will help you narrow it down to the actual species.

Look at the following features:

SHAPE

broadest at base   tapered to base   semi-tapering

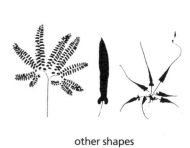

other shapes

DIVISION

undivided   once-divided   twice-divided   thrice-divided, lacy

no pinnae   pinnatifid: pinna   pinnate: pinna   pinna divided   pinnules divided
connected at base   separate at base

PINNA AND PINNULE MARGINS

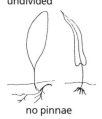

auricled:
ear-like lobe
at base

smooth   toothed   lobed

GROWTH FORM

lax

erect

clustered

scattered

STALK AND AXIS

scaly

hairy

woolly

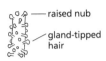
— raised nub
— gland-tipped hair

glandular

smooth   coloured   winged axis

ATTACHMENT OF PINNAE

opposite pinnae

alternate pinnae

sessile

stalked

pinnae or blade

REPRODUCTIVE FEATURES

The pattern on the underside of the spore-bearing fronds differs for each species and is one of the best ways to confirm the identification.

marginal

oblique or chevron

lined along margins

along midvein

scattered

VEGETATIVE REPRODUCTION

shoots from rhizomes
eg. Hay Scented fern

new plants from leaf tips
eg. Walking fern

bulblets which drop off and take root,
eg. Bulblet fern

Ferns Grouped for Identification Purposes

A. Where to Look for Ferns

Ferns listed by their most likely habitat.

Some ferns are found in more than one type of environment and are listed accordingly.

1(a) Mature Shady Woodland (unpastured for the last 50 years)

Cut Leaved Grape Fern	Rattlesnake Fern	Least Grape Fern
Daisy Leaf Grape Fern	Leathery Grape Fern	Maidenhair Fern
Hay Scented Fern	Northern Beech Fern	Hart's Tongue Fern
Lady Fern	Silvery Glade Fern	Narrow Leaved Glade Fern
Oak Fern	Bulblet Fern	Mackay's Fragile Fern
Intermediate Wood Fern	Spinulose Wood Fern	Clinton's Wood Fern
Goldie's Wood Fern	Male Fern	Marginal Shield Fern
Christmas Fern	Northern Holly Fern	

1(b) Mature Shady Woodland with Large Dolostone Boulders

Walking Fern	Maidenhair Spleenwort	Fragile Fern
Polypody		

2. Damp Rich Soil with Open Shade

Blunt Lobed Grape Fern	Cinnamon Fern	Interrupted Fern
Royal Fern	New York Fern	Marsh Fern
Ostrich Fern	Sensitive Fern	Silvery Glade Fern
Lady Fern	Oak Fern	Bulblet Fern
Spinulose Wood Fern	Crested Shield Fern	Clinton's Wood Fern

3. Dolostone Crevices, Rockface Dolostone, Alvars, or Talus Rubble

Slender Cliffbrake	Smooth Cliffbrake	Purple Stemmed Cliffbrake
Ebony Spleenwort	Walking Fern	Wall Rue
Hart's Tongue Fern	Maidenhair Spleenwort	Green Spleenwort
Robert's Fern	Bulblet Fern	Laurentian Fragile Fern
Fragile Fern	Mackay's Fragile Fern	Oregon Woodsia
Intermediate Wood Fern	Male Fern	Marginal Shield Fern
Northern Holly Fern	Polypody	

4. Dry Abandoned Pastures, Old Sand Dunes, Reforestation Areas

Moonwort	Mingan Moonwort	Least Grape Fern
Daisy Leaf Grape Fern	Leathery Grape Fern	Northern Adder's Tongue
Bracken	Ebony Spleenwort	Lady Fern

5. Thin or Open Shade with Acidic Soil

Northern Beech Fern	Hay Scented Fern	Interrupted Fern
Cinnamon Fern	New York Fern	Marsh Fern
Ebony Spleenwort	Silvery Glade Fern	Lady Fern

6. Wet Swampy Areas or with Sphagnum

Sensitive Fern
New York Fern
Virginia Chain Fern
Lady Fern

Royal Fern
Marsh Fern
Cinnamon Fern
Bulblet Fern

Crested Shield Fern
Spinulose Wood Fern
Ostrich Fern

On hummocks — Intermediate Wood Fern — Oak Fern — Northern Beech Fern

B. Evergreen Ferns

In these species the fronds remain green through the winter.
Look for the remains of previous year's fronds in spring or early summer.

Cut Leaved Grape Fern
[f. Ternate Grape Fern]
Ebony Spleenwort
Hart's Tongue Fern
Evergreen Wood Fern
Christmas Fern
Clinton's Wood Fern

Blunt Lobed Grape Fern
Smooth Cliffbrake
Walking Fern
Green Spleenwort
Crested Shield Fern
Northern Holly Fern

Leathery Grape Fern
Purple Stemmed Cliffbrake
Wall Rue
Maidenhair Spleenwort
Marginal Shield Fern
Polypody

C. Frond Divisions

The terms "small," "medium," and "large" are comparative within the group. In general, small would be under 20 cm (8 inches), medium from 20 to 60 cm (8 inches–2 feet), and large over 60 cm (2 feet).

Warning: Very young ferns may not show the characteristics of the adult. For example, young or poorly developed Lady Fern will be only twice divided.

1. **Plants with the frond divided into a sterile blade and fertile spike**
   (usually only one frond). Sporangia in pinnae of fertile spike.
   a. Blade undivided   Adder's Tongue Fern (*Ophioglossum pusillum*)
   b. Blade divided   **Botrychium species**
   i. Once divided — Least Moonwort *B. simplex*
                          Moonwort *B. lunaria*
                          Mingan Moonwort *B. minganense*
   ii. Twice divided — Daisy Leaf Grape Fern *B. matricariifolium*
                          Blunt lobed Grape Fern *B. oneidense*
                          Ternate Grape Fern *B. dissectum* forma *obliquum*
   iii. Thrice divided — Rattlesnake Fern *Botrychium virginianum*
                          Leathery Grape Fern *B. multifidum*
                          Cut Leaved Grape Fern *B. dissectum* forma *dissectum*

2. **Plants with undivided blades** — sporangia in elongate strips on back of blade.
Hart's Tongue Fern (*Asplenium scolopendrium*)
Walking Fern (*Asplenium rhizophyllum*)

Note: In groups 3 to 5 the sporangia are found in a variety of positions, or on separate fertile fronds — see section D.

3. **Once divided — blade divided into simple pinnae.**
   a. Small   Maidenhair Spleenwort *Asplenium trichomanes*
               Green Spleenwort *Asplenium trichomanes-ramosum*
               Purple Stemmed Cliffbrake *Pellaea atropurpurea*
               Smooth Cliffbrake *Pellaea glabella*
               Note: In both *Pellaeas* the bottom pinnae may be divided again.

   b. Medium   Polypody *Polypodium virginianum*
               Ebony Spleenwort *Asplenium platyneuron*
               Sensitive Fern *Onoclea sensibilis*

   c. Large   Narrow Leaved Glade Fern *Diplazium pycnocarpon*
               Christmas Fern *Polystichum acrostichoides*
               Northern Holly Fern *Polystichum lonchitis*

4. **Twice divided — blade with the pinnae divided into pinnules.**
   a. Small   Wall Rue *Asplenium ruta-muraria*
               Slender Cliffbrake *Cryptogramma stelleri*
               Oregon Woodsia *Woodsia oregana*

   b. Medium   Long Beech Fern *Phegopteris connectilis*
               Marsh Fern *Thelypteris palustris*
               New York Fern *Thelypteris noveboracensis*
               Crested Shield Fern *Dryopteris cristata*
               Lady Fern *Athyrium filix-femina*

   c. Large   Clinton's Fern *Dryopteris clintoniana*
               Male Fern *Dryopteris filix-mas*
               Goldie's Fern *Dryopteris goldiana*
               Marginal Shield Fern *Dryopteris marginalis*
               Virginia Chain Fern *Woodwardia virginica*
               Silvery Glade Fern *Deparia acrostichoides*
               Ostrich Fern *Matteuccia struthiopteris*
               Cinnamon Fern *Osmunda cinnamomea*
               Interrupted Fern *Osmunda claytoniana*
               Royal Fern *Osmunda regalis*

5. Thrice Divided — blade with pinnae divided into pinnules and the pinnules lobed (not merely toothed).

    a. Stalk dividing into two semicircular sections pinnae on one side.
        Maidenhair Fern *Adiantum pedatum*

    b. Axis appearing to divide into three i.e. bottom pinnae very large

i. Small	Oak Fern *Gymnocarpium dryopteris*
	Robert's Fern *Gymnocarpium robertianum*
ii. Large	Bracken *Pteridium aquilinum*

    c. Axis undivided

i. Small	Fragile Fern *Cystopteris fragilis*
	Laurentian Fragile Fern *Cystopteris laurentiana*
	Mackay's Fragile Fern *Cystopteris tenuis*
ii. Medium	Bulblet Fern *Cystopteris bulbifera*
	Hay Scented Fern *Dennstaedtia punctilobula*
iii. Large	Lady Fern *Athyrium filix-femina*
	Evergreen Wood Fern *Dryopteris intermedia*
	Spinulose Wood Fern *Dryopteris carthusiana*

D. Situation of Sporangia

1. Plants with fertile fronds distinctly "different" in form from sterile fronds

a. Small	Slender Cliffbrake *Cryptogramma stelleri*
b. Medium	Sensitive Fern *Onoclea sensibilis*
c. Large	Ostrich Fern *Matteuccia struthiopteris*
	Cinnamon Fern *Osmunda cinnamomea*

2. Plants with a section of the fertile frond "different"

a. Medium	Christmas Fern *Polystichum acrostichoides*
b. Large	Interrupted Fern *Osmunda claytoniana*
	Royal Fern *Osmunda regalis*

3. Plants with fertile fronds not noticeably different in form from sterile fronds

    a. Sori along edges of pinnae or pinnules — under inrolled margins
        Family: Pteridaceae — Maiden Hair Fern and Cliffbrakes
           and Bracken — *Pteridium aquilinum*

b. Sori elongate on veins at back of blade, pinnae or pinnules
   Family: Aspleniaceae — Spleenworts

c. Sporangia in sori on back of pinnules i.e. "spore dots"
   i.   Elongate sori in chains parallel to midrib
           Virginia Chain Fern *Woodwardia virginica*

   ii.   Sori with kidney shaped indusium
           Wood Ferns, Shield Ferns *Dryopteris* species

   iii.   Sori with round indusium
           New York and Marsh Fern *Thelypteris* sp.
           Northern Holly Fern *Polystichum lonchitis*

   iv.   Indusium round, cup-shaped - on edges of pinnules
           Hay Scented Fern *Dennstaedtia punctilobula*

   v.   Indusium hood shaped — attached at the side
           Fragile Ferns, Bulblet Fern *Cystopteris* species

   vi.   Indusium made of narrow threads
           Oregon Woodsia *Woodsia oregana*

   vii.   Spore dots with no indusium (use magnifier)
           Oak Fern, Robert's Fern *Gymnocarpium* species
           Long Beech Fern *Phegopteris connectilis*
           Polypody (large, conspicuous) *Polypodium virginianum*

   viii.   Sori elongate — forming a chevron pattern on back of pinnule or
                   pinna
           Narrow Leaved Glade Fern *Diplazium pycnocarpon*
           Silvery Glade Fern *Deparia acrostichoides*

   ix.   Sori — more comma shaped forming a herring-bone pattern
           Lady Fern *Athyrium filix-femina*

Annotated Fern Species List

Note: The dimensions given in the following descriptions apply to local specimens. The width refers to the widest part of the frond. The sign ± indicates an average. Dimensions will vary according to habitat conditions. Plants growing in marginal conditions may be very much smaller than is indicated here. Metric measurements have been used exclusively in the systematic descriptions but imperial measures have been retained in the text for the benefit of those who are not comfortable with the metric system.

1 inch = 2.5 cm.   10 cm = 4 inches   30 cm = 1 foot

1 metre = 100 cm = 39 inches (i.e. approximately 1 yard)

ADDER'S TONGUE FAMILY
Ophioglossaceae

The scientific name for this family comes from the Greek words *ophis* meaning **snake** and *glossa* meaning **tongue**. The common poisonous snake in England is the Adder which is the source of the English common name. This group of ferns looks quite different from any other. A stalked frond (usually only one) arises from the underground rhizome. The stalk divides into a sterile blade and a fertile spike. They are considered to be of very ancient origin and have some primitive characteristics, but probably due to their softer tissue, they have left no fossil record. Of the five genera found world wide, two are represented in our area, *Botrychium* and *Ophioglossum*, the latter being somewhat less primitive than *Botrychium*. Many of the species in this group are inconspicuous and grow in grassy or disturbed areas where they are not easy to spot and therefore are often overlooked.

Genus:
Moonworts or Grape Ferns
Botrychium
The Latin word *botrys* means **bunch** and refers to the resemblance of the fertile spike, in many species, to a miniature bunch of grapes. This is also the origin of the English name Grape Fern. "Wort" was the Anglo-Saxon term used for plants with culinary or herbal uses, and these plants are referred to in old herbals, often being ascribed magical qualities such as making you invisible! A 1591 reference in the Oxford English Dictionary states that a horse riding over Moonwort would lose its shoes by nightfall! In this genus the sterile portion of the leaves is subdivided. Most species produce only one frond from a small, unbranched, upright rhizome. Eight species of the thirty found in North America are present in Grey and Bruce Counties. *Botrychium lunaria* has a worldwide distribution, *B. virginianum* occurs throughout the Americas and in Eurasia. It is also the most widely distributed *Botrychium* species in North America. *B. matricariifolium, B. multifidum* and *B. simplex* are found in Europe as well as North America. *B. minganense, B. oneidense,* and *B. dissectum* occur in North America, although the range of the latter ranges into the West Indies.

Genus
Adder's Tongue Fern
Ophioglossum
In this genus the sterile portion of the leaf is undivided and tongue shaped, hence the name **Adder's Tongue Fern**. Only one of seven *Ophioglossum* species in North America occurs in this area, but it is very closely related to *Ophioglossum vulgatum* which is found farther south in the United States. In older books the two species are lumped together. *O. pusillum* is found in eastern North America north of the southern boundary of the Wisconsin glaciation in the temperate forest region. It also occurs from British Columbia to northern California as far east as Idaho and Montana.

1. Moonwort
Botrychium lunaria (L.) Swartz

In order to spot this exceedingly rare, little fern a "hands and knees" search is required. The single, rather leathery, frond appears in the late spring. It usually has four or more pairs of half-moon or fan-shaped pinnae with wavy margins. They are closely spaced to overlapping and slightly stalked. The overall height is usually less than six inches (15 cm). The fertile portion produces sporangial clusters, starting just above the level of the top of the sterile portion. They are borne on short branches, tending to droop. The spores mature in summer and the leaves die off.

The most likely places to find this fern are dry, grassy meadows, abandoned pastures or areas of reforestation. The small size and short life of the frond probably cause this plant to be overlooked, but it has become very rare in recent decades and may no longer be present. Any discoveries of this species should be reported to the M.N.R.

DESCRIPTION
Growth Form: One frond. Stalk ± ½ total length. Height up to 20 cm.
Sterile portion: Blade dark green, leathery. Once divided, 3-6 pairs of fan shaped pinnae, usually overlapping. Up to 12 cm long.
Fertile portion: Once or twice divided. 3-4 pairs of reduced pinnae and a terminal pinna bearing sporangia. Up to 20 cm long.
Season: Frond appearing in spring, dying by summer. Spores mature late June to early July.

2. Mingan Moonwort
Botrychium minganense Victorin

This small grape fern is rare but can be found on old sand dunes back from the Lake Huron shoreline. It has also been recorded from pine reforestation areas in the Sauble Beach area. It is rare in Bruce County but has not yet been recorded from Grey. It was originally described by the great Canadian botanist Frère Marie-Victorin who collected it on the Mingan Islands in the lower St. Lawrence River.

In this area it is less than six inches (15 cm) in overall height and is distinguished from its close relative, *B. lunaria*, by its small size and yellowish green colour (as opposed to dark green). In addition, the compressed fan shaped pinnae are ascending rather than at right angles to the stalk and they do not overlap one another. Confusion is more likely between immature plants of this species and the Least (Dwarf) Grape Fern (*B. simplex*). Both species are exciting finds and new locations should be reported to the MNR office in Owen Sound as they are rare and little is known about their distribution.

DESCRIPTION
Growth Form: One frond. Stalk ± ⅓ total length.
Height up to 10 cm.
Sterile portion: Blade dull yellow-green. Once
divided, ± 5 pairs fan-shaped pinnae, edges
sub-entire. Rarely overlapping.
Fertile portion: Once divided ± twice length of sterile
portion. Tapering to apex. Sporangia borne on the
reduced pinnae.
Season: Frond appearing late spring to summer.
Spores mature in July.

3. Least Grape Fern
Dwarf Grape Fern
Botrychium simplex **Hitchcock**

Though not the rarest of our Grape Ferns, it is, nevertheless, a rare species and one of the most difficult to discover. Least Grape Fern hides among grasses and other plants, making a search on hands and knees a necessity. This small, pale green fern is smooth and fleshy and is most likely to be found in moist woodlands or on pasture edges, sometimes on rather poor soil. It is less likely to be seen near the dolostone of the escarpment as it requires some acidity.

Overall, it is between two to five inches (5-13 cm) tall. The sterile frond is about one and a half inches (4 cm) long with the shape varying from almost entire to lobed or once divided into a few pairs of lobed pinnae. The position on the stem also varies: sometimes it clasps the stem and sometimes it is stalked. The fertile portion rises above the blade and the prominent sporangia are widely spaced. The base of the stalk has a sheath just above the fleshy root.

These plants appear in early spring and disappear in early summer. Many of them are so small that they can be found only by removing the leaf litter covering them. Least Grape Fern has been found in isolated spots from southern Grey County to the Bruce Peninsula, but it is probably much more frequent than it appears from the records because of the difficulty of seeing it and because of the short life span. It is also difficult to distinguish from immature plants of *B. minganense*.

DESCRIPTION
Growth Form: One frond. Stalk ± ⅔ height. Overall height to 13 cm.
Sterile portion: Blade bright to whitish green. ± 3 cm long. Somewhat fleshy. Once divided into shallow, blunt pinnae ± 3 pairs.
Fertile portion: Usually once divided, 2-3 pairs of reduced pinnae each bearing a few sporangia.
Season: Frond appears in spring. Spores mature in June.

29

4. Daisy Leaf Grape Fern
Botrychium matricariifolium (Döll.) A.

In common with most of the Grape Ferns, this species is small and hard to find in Grey and Bruce Counties. Plants appear in early spring and wither by midsummer. They may be as much as one foot (30 cm) in height but are generally smaller. The fronds will be found standing erect with the single blade high up on the fleshy stalk, not far below the fertile portion. The shape of the blade is variable but it is short stalked and usually has four or five pairs of pinnae, the lower ones being quite deeply lobed, progressively less lobed higher up the blade. They are spaced apart from one another, and the leaf is best described as daisy-like. The spores mature in late June or July and the fertile segment withers shortly after.

This species is found away from the escarpment as it requires more acid conditions. The edges of old sandy and sterile fields, dry woods or even moist cedar woods are places to look for it, though it is considered to be rare in this area.

DESCRIPTION
Growth Form: One frond. Stalk ± ⅔ total length. Height up to 30 cm.
Sterile portion: Blade dull green. Once or twice divided, ± 5 pairs blunt pinnae, usually with blunt teeth.
Fertile portion: Twice length of sterile portion with long stalk. Irregularly, once or twice divided, lower segments longer. Sporangia borne on the reduced pinnae.
Season: Frond appearing late spring, dying by late summer.

5. Cut Leaved or Dissected Grape Fern

Botrychium dissectum Sprengel

[*Botrychium obliquum* Muhl. – Ternate Grape Fern]

This fern produces two leaf forms, one in which the blades are obviously thrice divided and lacy (forma *dissectum*) and another in which they are much less finely divided with broader pinnules and lobes (forma *obliquum* or Ternate Grape Fern). Both forms are often found growing together. One strange feature of this fern is that new growth does not appear until August or September so that the sterile frond has a fresh green appearance in the early fall but darkens as the frost comes, giving a bronze-green colour throughout the following spring and summer.

Dissected Grape Fern is of medium size. The triangular blades are four to six inches (10-15 cm) long and often parallel to the ground. The terminal lobes are larger and more elongated than those of *B. multifidum*, for which it is often mistaken. The fertile portion appears in September, branching from the main stalk close to ground level. The yellow-green sporangia form on branches on the upper part of the six to eight inch (15-20 cm) long, erect stalk. The yellow-green spores mature in early fall, the fertile stalk withers and the sterile frond over-winters.

This is not a common fern, but both forms are found in Grey and Bruce Counties, usually in upland hardwood forest and sometimes in poor soil in the open. It may be seen right on the edge of the compacted Bruce Trail, most notably in upland hardwoods along the Hope Bay "Pothole" section. It is also found in the Woodford Conservation Area and Bayview Nature Reserve in Sydenham Township.

DESCRIPTION

Growth Form: One frond. Stalk ± ⅔ total length. Height including fertile spike ± 20 cm.

Sterile portion: Blade shiny green, turning bronze in winter, triangular, thrice divided. Margins very finely divided. 3-20 cm long 4-15 cm wide.

Fertile portion: Rises above sterile portion, top third once (or twice) divided, sporangia in a double row on reduced pinnae.

Roots: Thick and dark grey to brown.

Season: Frond overwintering, new frond in late August or September. Fertile portion best seen in late September or early October.

Ternate
f. obliquum

Cut Leaved
f. dissectum

6. Blunt Lobed Grape Fern
Botrychium oneidense (Gilbert) House

The Blunt Lobed Grape Fern has fronds eight to twelve inches (20-30 cm) tall when fertile. The sterile blade is triangular and the pinnae are oval and blunt tipped. New fronds appear in late summer and remain evergreen (not turning bronze) throughout the winter and into the following summer. Most plants found in this area are without fertile segments.

Botrychium oneidense may be difficult to distinguish from young plants of both *B. dissectum* and *B. multifidum*. *B. oneidense* has fewer, large segments well spaced apart. The length of the stalk in the sterile segment in the Blunt Lobed Grape Fern is much longer, when compared to the length of the blade, than it is in the other two species. The fertile segment also is more than double the length of the sterile segment and is comparatively much shorter in *B.dissectum* and *B. multifidum*.

This species is found in low, wet, shady woods and swamps with acid soil, never in open fields. It appears to be rare in Grey and Bruce Counties as well as in Ontario generally, but this may be due partly to its recent recognition as a separate species and the difficulty of identification.

DESCRIPTION
Growth Form: One frond branching just above the ground. Height with fertile segment ± 25 cm.
Sterile portion: Blade dull bluish green somewhat leathery. Three divisions with 1-3 pairs of broad ovate subdivisions, merely lobed at apex. Margins wavy to denticulate.
Fertile Portion: Long stem. 2 to 3 x length of blade. Once (twice) divided, sporangia borne in two rows on reduced pinnae.
Roots: Slender and buffy in colour compared to *B. dissectum*.
Season: Frond overwinters. Green. Spores mature in last half of September.

7. Leathery Grape Fern
Botrychium multifidum (S.G.Gmelin) Ruprecht

This fern derives its name from the leathery texture of the overwintering fronds. Frequently, it has more than one, overlapping, sterile leaf, unlike other species of *Botrychium* which most often have a single leaf. New bright green fronds appear among the yellowed remnants of the old fronds in July or August. The fertile and sterile segments branch off near ground level. Later in the season the blade becomes darker. The toughness of the pencil-thick basal stalk sometimes enables the fronds (which may be up to one foot (30 cm) in overall height) to stand erect throughout the winter. The three main divisions of the broadly triangular blade are crowded and overlapping, deeply subdivided and varying from slightly pointed oval to fan shaped. The terminal lobe is much smaller than in *B. dissectum*. This species sometimes has several crowded fronds.

It may be found in exposed meadows, on grassy hillsides, in sandy areas that are open or deciduous forested, and in other non-swampy deciduous forests, but overall it is rare.

DESCRIPTION
Growth Form: Stalk short 2-5 cm. Overall height ± 25 cm.
Sterile portion: Blade shiny green, leathery. Stem length very short. Blade ±10 cm long, width ± 10 cm. Thrice divided, lobes obtuse to subacute, somewhat overlapping.
Fertile portion: Twice length of sterile portion, twice divided, sporangia borne on the reduced pinnae.
Season: Frond evergreen. Spores mature in early fall.

8. Rattlesnake Fern
Botrychium virginianum (L.) Swartz

This triune fern has each section thrice divided and is, by far, the most common and the largest of our *Botrychium* species. The fertile spike is thought to resemble the rattle of a rattlesnake. One is likely to come across it in shady woods anywhere in the area. It is more often found in deciduous rather than coniferous woods, mostly in uplands but normally not in swamps.

This is a succulent fern with soft, juicy tissue in those parts above the ground. It arises from a small, deep, erect rhizome with spreading, fleshy roots. The base of the stalk, just below the surface, is expanded into a sheath which encloses three or four leaf buds, one for each succeeding year. Rattlesnake Fern does not produce fiddleheads. The fronds form underground during late winter and early spring. When the time is ripe, fully formed, compactly folded leaves are pulled upward and open only when completely above ground. The plume-like fertile portion springs stiffly, on a long stalk, from the base of the sterile blade which is about halfway up the main stalk. These ferns are peculiar in that they will produce spores no matter how small and impoverished the plant. Shortly after discharging its cream-coloured spores, like a puff of smoke, the fertile portion of the leaf withers. The sterile portion, if conditions are favourable, will remain green and juicy all summer, attracting its worst enemies — slugs and snails. Early American Indians used the roots in a poultice as a remedy for the bites of poisonous snakes. A decoction of the root was also made and sprinkled on the ground around dwellings to ward off snakes.

DESCRIPTION
Growth Form: One frond. Stalk ± half total length. Overall height ± 42 cm.
Sterile portion: Blade pale green, triangular, ± 20 cm x 25 cm, thrice divided, lobes toothed.
Fertile portion: Stem ⅔ length. Twice divided. Sporangia borne on small pinnae.
Season: Frond appears mid-spring, fertile portion later. Dies in mid-summer. Spores mature in July.

9. Northern Adder's Tongue
Ophioglossum pusillum **Rafinesque**

The Adder's Tongue is a strange, unfernlike, little plant with its fleshy stalk tipped by a single, fertile spike, withering by mid-summer. It is supposed to resemble a snake's tongue. The oval leaf portion of the frond is about three inches (7 cm) long with tiny, netted veins forming a distinct, chainlike pattern. This smooth, succulent, grass-green leaf can hide in the grasses of abandoned fields on former sandy shorelines and never be noticed. In damp places farther south in Ontario, it may stretch to one foot (30 cm) in overall height but is usually much shorter. It is said to be widely distributed, but since is difficult to find, is seldom recorded; nevertheless it is rare in Grey and Bruce.

The fertile section consists of two rows of sporangia embedded in the upper part of the stalk. The plants spread out from fleshy roots, accounting for its habit of growing in little patches. The leaf for the succeeding year arises from a bud within the sheath at the base of the stalk. Historically, Adder's Tongue was used for healing wounds in the form of an ointment, known as "Green Oil of Charity." It was also used by artists as a green pigment.

DESCRIPTION
Growth Form: One frond arising from the rhizome.
Sterile portion: Blade dull pale green. Long oval (lanceolate to ovate), may be folded. ± 8 cm long, ± 2.5 cm wide
Fertile portion: Twice length of blade. Unbranched, terminating in a double row of sporangia, 10-40 pairs, with a sterile tip.
Season: Frond appearing in late spring. Not overwintering.

"FLOWERING" FERN FAMILY
Osmundaceae

Worldwide, there are three genera in this family only one of which, **Osmunda**, is represented in Ontario.

Genus:
Osmunda

The origin of the name *Osmunda* is something of a mystery. Linnaeus named the genus in 1753, but he must have been aware that it had been applied to ferns in western Europe as far back as the 11th century, the Anglo-French word being **Osmonde** or **Osmunda** in mediaeval Latin (Oxford English Dictionary). There are many names of Norse and Saxon origin using the syllable **Os** which means "god." Osmond is a personal name that means god-protector. The suggestion, dating back to the last century, that *Osmunda* comes from the name Osmunder, Saxon God of War (Thor by another name) cannot be confirmed. There was no such god (pers. comm. Dr. Richard Perkins, Univ. of London, U.K.).

At first, *Osmunda* appears to have been a general term for ferns, as it was also used to refer to the Male Fern. By 1600, the Royal Fern was known as Osmund Royal. Prior to that, it was sometimes called Osmund the Waterman, which refers to its preference for watery habitats. The term "flowering fern" was coined in England because in the Royal Fern (their only representative of this group), the fertile portion of the frond sprouts from the top of the leafy part and looks not unlike a flower such as Meadow Rue. Of our three species, Royal Fern is the most widespread, being found in a number of varieties worldwide; var.*spectabilis* is the North American version.

Cinnamon Fern occurs in temperate areas of eastern North America through Texas and Mexico to Central America, the Caribbean Islands and South America. Curiously, Cinnamon Fern is absent from the rest of the world including western North America. Interrupted Fern occurs in North America from Newfoundland through Ontario to Minnesota and south through the Appalachian States as far south and west as Arkansas. It also occurs in eastern Asia. It ranges farther north in Ontario than Cinnamon Fern and Royal Fern, being quite common in the Thunder Bay area where the other two are not found.

10. Cinnamon Fern
Osmunda cinnamomea L.

Cinnamon Fern derives both its Latin and English names from the colour of the mature fertile frond which, as the spores develop, looks like a wand covered with powdered cinnamon. The pinnae are small and dark brown, and the frond is clad in cinnamon brown wool. The fertile fronds appear first and wither early in the season. The sterile fronds surrounding them are large and green, producing no spores and persisting until the first frosts. Some plants produce only sterile fronds. They can be distinguished from infertile fronds of the Interrupted Fern in three ways. First, they are a yellower green. Second, there are tufts of cinnamon wool at the junction of the pinnae and axis, not occurring in the Interrupted Fern. Third, the outline of the frond is more pointed in the Cinnamon Fern. The contrast in colour between the fertile and sterile fronds makes this fern an interesting subject for photography in the latter part of June when it is in prime colour, before the fertile fronds wither. The fiddleheads, too, as they emerge in spring, have spectacular highlights from the extended white hairs in which they are cloaked.

This is one of the larger, more luxuriant ferns but not one of the most common overall in Grey and Bruce Counties, probably because our soils tend not to be acidic. It may be fairly plentiful in suitable habitats such as acidic, swampy woods (most often mixed or coniferous) but will not generally be found on the dolostone of the escarpment. Cinnamon Fern also occurs in open bogs where the fronds will be lighter green. This fern is less common on the Bruce Peninsula than farther south. It is widespread throughout southern Ontario and is found as far north as Lake Superior.

DESCRIPTION:
Growth Form: Clumps of fronds arising from a stout, central rhizome with persistent leaf stalks. Often partly above ground. Fertile fronds at centre.
Sterile Frond: Length up to ± 1.0 m. Width ± 25 cm.
Blade: Twice divided, deep, slightly yellowish green.
Pinnae: 15-25 pairs, lanceolate tapering, sharply pointed, subopposite, length 6-11 cm.
Pinnules: 15-20 pairs, oblong, wider at base, not cut to mid-vein, rounded apex. Narrower than Interrupted Fern. Alternate.
Axis: Smooth, green, slightly grooved in front, cinnamon woolly early in season. Tufts persisting at junction with pinnae.
Stalk: ¼ to ⅓ total length of frond, also woolly early in season. Pinkish brown.
Fertile Frond: Slightly longer than sterile frond, wand-like, pinnules not expanded.
Appear first, wither in early July. Some clumps produce no fertile fronds.
Sporangia: Densely packed clusters on fertile pinnules. Colour dark green soon changing to brown.
Season: Fiddleheads appear in May; large, covered with silvery hairs. Fronds not evergreen; gone by November.

11. Interrupted Fern
Osmunda claytoniana L.

This large fern with its gracefully curving fronds arising in a cluster from the central rhizome, is unusual in that a group of fertile pinnae forms about half way down the axis. This is the source of its English name. The outer fronds are sterile. The central fronds, which are more erect, produce from three to five pairs of fertile pinnae. These are dark green at first, but as the sporangia ripen, they become dark brown, a process which takes about two weeks.

Moist conditions are required, but it will not grow in really wet areas or in areas of dense shade. It is commonly found in moist woodlands (especially mixed forest), often along woodland roads and in open thickets, sometimes also on lake or stream banks, well above water level. Subacid soil is preferred although it is a little more tolerant of calcium than Cinnamon Fern. While found in suitable habitat throughout Bruce and Grey Counties, it is less common in Grey County. It seems to be most frequent in Greenock Swamp and from Sauble Falls to Pike Bay. It is easy to see along the roadsides in the Oliphant to Red Bay area. However, it is not a common fern in this area, being very much less common than Cinnamon Fern, possibly because places that are not too wet for it tend to be insufficiently acidic.

DESCRIPTION:

Growth Form: Fronds arising in a cluster from the stout, central rhizome which is covered with persistent leaf bases and fibrous roots.
Sterile Frond: Length ± 80cm. Width ± 25 cm.
Blade: Twice divided. Slightly bluish green.
Pinnae: Lanceolate. Sessile.
Pinnules: Oblong, not cut to mid-vein. Rounded.
Axis: Glabrous. Straw colour with dark groove. (No tufts of hairs at junction with pinnae.)
Stalk: About ⅓ length of frond. Covered with loose, cottony hairs in spring soon becoming glabrous. Tan colour.
Fertile Frond: Slightly taller than sterile frond. 3-5 pairs of reduced, fertile pinnae in middle of the frond. Many clumps produce no fertile fronds.
Sporangia: Long-stalked, form in clusters on fertile pinnules. Green changing to dark brown on maturity.
Season: Fiddleheads appear in May. Not evergreen.

12. Royal Fern
[Regal Fern, Flowering Fern]
Osmunda regalis L. var. *spectabilis* (Wildenow) A. Gray

Both the Latin and English names refer to the majesty of this large and graceful fern. In spring it is most colourful, both the fiddleheads and young fronds showing shades of orange to wine-red blended with greenish brown. The blades are so deeply divided that the pinnae look almost like a Mountain Ash leaf. Fertile leaves produce several pairs of fertile pinnae at the apex of the blade. Of the three species of *Osmunda*, the Royal Fern is best adapted to very wet areas.

It is found throughout Grey and Bruce Counties in all kinds of wet situations — swamps, fens, lake margins and along river banks. Royal Fern is particularly abundant in the Greenock Swamp and through the drainage basin of the Rankin River. In the Red Bay area it can be viewed along the roadside. It probably has less tolerance for acid conditions than the other two ferns in this group and needs somewhat less shade. This is certainly the most common species of this genus in Grey and Bruce Counties.

DESCRIPTION:
Growth Form: Fronds clustered, arising from a stout, erect rhizome covered with persistent leaf bases and fibrous roots.
Sterile Frond: 60-140 cm long, 30-50 cm wide.
Blade: Twice divided. Pale green, slightly translucent.
Pinnae: Length ± 20 cm. Subopposite.
Pinnules: ± Sessile. Alternate. Oblong — base rounded and oblique, apex bluntly pointed. Serrulate. Distinct "feathered" veining.
Axis: Glabrous. Straw coloured. Grooved.
Stalk: ± ½ length of frond. Glabrous. Straw coloured, reddish at base. Rounded.
Fertile Frond: Several pairs of reduced, fertile pinnae at apex. Greenish to brownish at maturity. Pinnae in lower 3/4 fully expanded and sterile.
Sporangia: Stalked. Form in clusters on fertile pinnae.
Season: Fiddleheads appear in May. The fronds wither by early fall.

MAIDENHAIR FERN FAMILY
Pteridaceae

The family name is derived from *Pteris*, the Greek word for fern. There are approximately 20 genera worldwide with 1000 species. Only three genera are represented in Bruce and Grey Counties. The attractive Maidenhair Fern *Adiantum pedatum* is most often seen. In older books the family name derived from this genus, i.e. **Adiantaceae**, may be found.

Genus:
Adiantum

The Greek word *adiantos* means unwetted and refers to the smooth surface of these ferns which readily sheds water. The characteristic feature of this genus is that the edges of the lobes curl under and protect the clusters of sporangia which form in the groove. This is known as a "false indusium" as opposed to a true indusium which is a separate protective structure formed on the leaf surface. Our only species — Maidenhair Fern — is found thoughout temperate Eastern North America but it is closely related to two other species which occur in Western North America and Asia.

Genus:
Cryptogramma

Cryptogramma is derived from the Greek words *cryptos* meaning "hidden" and *gramme* meaning "line," referring to the line of sporangia hidden in the rolled-under edges of the leaflets of the fertile fronds. It is also characteristic of this group that the sterile frond is clearly different from the fertile frond. The word **brake** is an old English term, related to the word **bracken**, once used to refer to ferns in general but now only applied to the ferns in this group which are found on rocks or cliffs. Our only species — Slender Cliffbrake — is found in Eastern and Western North America but not in the centre, a distribution related to its habitat requirements.

Genus:
Pellaea

The Greek word *pellos* means "dark" and was probably used for this genus in reference to the characteristic dark stems, although the leaves too tend to be a rather dark bluish gray. The old English word brake is often used in the common names of this group. The difference between the sterile and fertile fronds is not so marked as in *Cryptogramma*. Of the two species found in our area, Smooth Cliffbrake is a North American endemic, the main distribution being in temperate, eastern North America. Purple Stemmed Cliffbrake, while a rare plant in Canada, has a more extensive distribution extending into the southwestern United States and Mexico as far south as Guatemala.

13. (Northern) Maidenhair Fern
Adiantum pedatum L.

If Nels Maher, local fern enthusiast, had to choose his favourite woodlot fern it would be this dainty and graceful plant which shimmers in the slightest breeze. In early spring it changes almost daily; first the transparent fiddleheads appear, full of warm colours. They uncoil rapidly and reach almost their full size in a week to ten days. The wiry, polished, purple-brown stalk divides and spreads forming a horseshoe-shaped frond with about twelve horizontal pinnae fanning out nearly parallel to the ground, a growth form which makes this fern quite unmistakable. When the leaves are mature the sori are situated in the reflexed lobe margins on the outer, curved side of the pinnule, forming a line of dashes.

Maidenhair ferns, in favourable habitat, spread slowly outwards from the original plant, forming a lush colony which may be up to forty square feet (4 m²). The Massie cross-country ski area has some of the finest specimens in Grey County. The rocky slopes of Sugar Maple woods, with rich soil, are its favourite haunt. In southern Bruce County it is found mostly in rich, upland hardwood stands away from the Lake Huron shore. It also occurs in quantity on the Bruce Peninsula, particularly toward the escarpment as far north as Lion's Head, but becomes sporadic farther up the Peninsula. In our area an abundance of Maidenhair Fern (perhaps more than any other species) tends to indicate rich non-sandy soil.

The pattern of the leaf is reminiscent of an animal's paw; the branching of the axis also suggests the anatomical subdivision of the bones of the foot, giving rise to the specific name *pedatum* which means footed in Latin. The leaves of *Adiantum pedatum* were used by Indians in the form of a medicinal tea used to treat consumption, coughs and other respiratory diseases. The dark shiny stems were also valued for their decorative qualities in basket making.

DESCRIPTION
Growth Form: Fronds growing in colonies from horizontal rhizomes.
Fronds: Length ± 55 cm, width ± 40 cm. Stalk ± 38 cm, purple-brown. Forked, arching back, twice divided, producing a fan-shaped blade.
Pinnae: Oblong, formed on one side of stem.
Pinnules: Stalked, oblong, lobed on one side.
Sporangia: In sori formed in reflexed lobe tip. Spores mature late summer.
Season: Fiddleheads appear from late April. Not evergreen.

14. Slender Cliffbrake
(Rockbrake)
Cryptogramma stelleri **(S.M.Gmelin) Prantl** *in* **Engler**

Slender Cliffbrake is a small dimorphic fern which is very habitat specific. Unless you look for it on a dolostone ledge or shelf which is dripping with condensation or seepage, your chance of finding it is very slim. This fern appears in spring and grows from three to six inches (7-15 cm) in height. It dies back in August or even earlier in a dry summer. The best time to hunt for it is in June or July. It is delicate and will grow only where it is fully protected from both wind and sun. The sterile frond could be mistaken for a small immature Fragile or Bulblet Fern as it is frail, ovate and similarly divided, but most often, the slightly taller, fertile frond with its unique unfernlike shape is growing right beside it. The sori are marginal along the ends of the veins at the edges of the fertile pinnules. These edges curve over to form a false indusium, often meeting in the middle and totally enclosing the back of the pinnule causing it to look very slender, hence its common name.

Some of the best stations for this fern are Feversham Gorge, Rocky Saugeen River near Highway 6, Owen Sound Waterworks springs, Spirit Rock near Wiarton, Sydney Bay in Cape Croker Indian Reserve and Crane River. On the Bruce Peninsula from Owen Sound northwards, it is almost restricted to the Niagara Escarpment including the Tobermory Islands.

DESCRIPTION

Growth Form: Fronds scattered along slender creeping, branched rhizome, deep in the rock.
Sterile Frond: Twice divided. Rather limp. Overall length 3-15 cm, width ± 4 cm . Stalk pale green to mauve, ½ length of frond.
Pinnae: Ovate.
Pinnules Blunt and rounded, may be shallowly lobed again at base of pinna.
Fertile Frond: Twice divided at base. Once divided near apex. Erect. Height ± 12 cm.
Pinnae: Irregularly divided with 2-3 pairs of blunt, elongate pinnules. Upper leaflets undivided.
Sporangia: In sori formed under the rolled-under edges of the fertile pinnules, forming a line round the margin.
Season: Fronds appear in spring and die by late summer. Rhizomes are biennial.

15. Purple Stemmed Cliffbrake
Pellaea atropurpurea (L.) **Link**

Purple Stemmed Cliffbrake has been found only on dolostone, in our area, in four township units on the Bruce Peninsula south to the Cape Croker Reserve. It grows on outcrops, alvars, or much less often, at or below the escarpment cliffs. It springs from dry crevices in the rock, with little or no soil and full exposure to the sun. It appears much later in the spring than most ferns, and in very dry summers may disappear altogether, only to re-emerge the following year when conditions become more favourable. The name Purple Stemmed is supposed to differentiate this species from the Smooth Cliffbrake, but as the latter has a reddish-brown stem this may be hard to see in the field. However, the Smooth Cliffbrake, as its name suggests, has a smooth and shining stalk, whereas the Purple Stemmed Cliffbrake has a stalk which is duller, with tiny bristly hairs which are hard to see but obvious when you run your fingers down it.

The widely spaced, narrow, pointed, bluish-green pinnae are evergreen. The lower ones are divided into three pinnules. The sterile fronds are shorter and have more oval segments. The sporangia are borne in the curled-under margins of the fertile segments. They are bright brown at maturity. In addition to the differences in stem characteristics, Purple Stemmed Cliffbrake tends to be taller, more erect and with fewer fronds than Smooth Cliffbrake. There is no noticeable difference between the fertile and sterile fronds of the latter which grows only on cliffs. If these differences are kept in mind, it is not too difficult to separate these two species.

Purple Stemmed Cliffbrake is a rare species in our area and is listed as a provincially rare plant in Ontario.

DESCRIPTION

Growth Form: Fronds arising stiffly from an erect, compact rhizome ± 7 mm diameter.
Sterile Frond: Stalks purplish-brown, bristly hairs. Blade bluish-green. Somewhat leathery. Partly twice divided. Overall length ± 20 cm, width ± 4 cm. Stalk ⅓ length of frond.
Pinnae: Ovate, stalked. Lower ones may be divided into pinnules.
Fertile Frond: Dark blue-green. Partly twice divided. Height ± 20 cm, width 4 cm.
Pinnae: Linear oblong, stalked. Lower ones divided into three pinnules.
Sporangia: In sori formed under the in-rolled margins of the fertile pinnules.
Season: Evergreen.

16. Smooth Cliffbrake
Pellaea glabella Mett.

Although not listed in the Peterson Field Guide, this fern is quite common in the Owen Sound area where it finds a precarious foothold on perpendicular dolostone cliffs. Its fronds sprawl rather ungracefully, but it compensates for this with its fascinating bluish-grayish-green colour and the polished reddish-brown, wiry stalks. Its habit of locating high out of reach on the escarpment makes it difficult to photograph unless you are lucky enough to spot an exceptionally low one.

The sporangia form under the recurved margins of the pinnules of the fertile fronds which are slightly longer than the sterile fronds. The size of the plants varies greatly from two or three fronds only 3 or 4 inches (7-10 cm) long to massive clumps the size of a cabbage head with fronds 10 inches (25 cm) long. Smooth Cliffbrake seems to do best in full sunlight, but its very fine roots penetrate deep into tiny cracks in the limestone to obtain the moisture it requires. The large, old clumps contain the accumulation of many years' growth of the brittle, shiny, smooth stalks from which it gets its name. It differs in this respect from the Purple Stemmed Cliffbrake which has little hairs and barbs on the stalks. The tough evergreen fronds and unusual coloration make it seem very "unfernlike." When looking for it with binoculars on the high face of the escarpment, you will soon discover that there are few ferns or any other vegetation growing around it in this inhospitable habitat. There are places around Owen Sound where thirty or so plants can be counted from one spot. It occurs in other places along the length of the escarpment but becomes sparse north of Lion's Head. Where dolostone outcrops away from the escarpment, Smooth Cliffbrake has been found — for example along the Rocky Saugeen River.

DESCRIPTION

Growth Form: Fronds arising from an erect, compact rhizome.

Fronds: Overall length ± 15 cm, width ± 4 cm. Stalks ⅓ length of frond, reddish-brown, hairless and glossy. Blade bluish-green. Stiff and leathery, evergreen. Once/twice divided.

Pinnae: Oblong-lanceolate, sessile. Lowest ones often divided into three pinnules.

Sporangia: In sori formed under inrolled margins of fertile pinnae.

Season: Evergreen.

Bracken Family
Dennstaedtiaceae

Until recently, the members of this family were included in the Pteridaceae. Studies have now shown differences, one of the main ones being that species in this group generally form a true indusium. The name is derived from that of the German botanist A.W. Dennstaedt who was influential in plant taxonomy at the beginning of the 19th century. There are about twenty genera worldwide, of which only two are represented in our area, by the ubiquitous Bracken and the Eastern North American endemic, Hay Scented Fern.

Genus:
Cuplet Ferns
Dennstaedtia

The name of this genus has the same derivation as that of the family. The English name "cuplet fern" refers to the characteristic "cups" containing the sporangia which are formed from the indusium and the tip of the pinna lobe.

Genus:
Pteridium

As with the family Pteridaceae, this genus derives its name from the Greek word *pteris* meaning fern. Bracken is the sole species in this genus, although many subspecies have been described. It is, undoubtedly, the best known fern in the world. One reason for this is that it is extremely well adapted to human activities, especially clearing forest, introducing grazing animals and burning pasture. There is some evidence that the spores, at least, are carcinogenic. There is a high incidence of stomach cancer in Wales where Bracken on the cleared uplands surrounds the reservoirs for the water supply. The fiddleheads are eaten in some countries but this is probably unsafe. It is also dangerous to livestock if eaten in large quantities.

17. Hay Scented Fern
Dennstaedtia punctilobula (Michaux) T. Moore

Hay Scented Fern is very common in New England where it invades rocky pastures and is known as "Boulder Fern." Grey and Bruce Counties are on the western fringe of its range. In addition, it is considered to be acid loving, so finding this fern in this area is a challenge. It has been located in only six townships so far. Hay Scented Fern grows in shade in our area. It is known to tolerate prolonged wet or dry seasons. The fronds are two feet (60 cm) long and individually (but closely) spaced along a shallow, creeping and branching rhizome that soon becomes matted, excluding all other plants. From a distance, Hay Scented Fern can be confused with Lady Fern or Silvery Glade Fern, but it may be recognised by the sweet scent of new mown hay which is released when the blade is crushed between the fingers or when it is dried. The small sori form at the indents of the pinnule lobe margins. The cup-shaped indusia open at the top and contain few sporangia. The yellow-green frond is very glandular with prominent bristly hairs on the lower surface. The lance-shaped blade becomes widest near the base. The stalk is one quarter the length of the frond, bristly throughout, almost black at the base, becoming light or reddish-brown above.

This thrice divided lacy fern with its fronds gracefully arching in many directions dies back early, sometimes turning a creamy white before becoming light brown. Although this fern is an acidophile, in this area it occurs mostly in calcareous upland deciduous forests. It is rare overall in both counties and extremely rare on the Bruce Peninsula. The single known location on the Bruce is at Hope Bay where it is growing in what appears to be a glacial deposit between two limestone ridges, along with another of our rather uncommon species, New York Fern.
It has also been found in Greenock Swamp.

DESCRIPTION
Growth Form: Fronds arise from slender, creeping, branched rhizomes. Often in large patches.
Fronds: Overall length 30-70 cm , width ± 15 cm. Stalks ⅓ or less length of frond, dull, pale brown, minutely glandular. Blade lanceolate, slightly narrowed at base. Thrice divided. Pale green. Soft silver-grey hairs on both surfaces. Characteristic scent of hay when crushed.
Pinnae: Lanceolate. ± 20 pairs.
Pinnules: Oblong, divided into sharp lobes.
Sporangia: In tiny sori formed in cup-shaped indusia on the upper margin of the lobes.
Season: Fronds appear in spring, fading by early fall.

18. (Eastern) Bracken

Pteridium aquilinum (L.) Kuhn ssp. *latiusculum* (Desvaux) Underwood

Bracken is a very common fern in Grey and Bruce Counties. Wherever it gains a foothold it is prone to take over as a weed. The tough rhizomes branching extensively deep in the ground are very difficult to eliminate completely. For this reason, it quickly recovers after a forest fire. Nevertheless, it provides shelter for many small animals and even other species of fern, such as *Botrychium* which may be found beneath its coarse, triangular fronds. Bracken exhibits some preference for poor soils, growing in open woods, whether damp or dry, and on sandy hillsides where it forms dense masses. Walking through such a patch with tough stems three to four feet (90 - 120 cm) high is very strenuous exercise. The sporangia form under the recurved edge of the pinnules, making them very inconspicuous.

Folklore and fable had many references to Bracken. In the past it was used in mattresses to prevent rickets, in pillows for the relief of asthma, and in many old time medicines. Indians often wore it over their heads to discourage blackflies. Bracken is eaten in the Orient, contributing to the high incidence of stomach cancer there, as it is carcinogenic (Hodge, 1973). However, most of our native ferns are likely to be toxic to some degree and should not be eaten in any quantity.

DESCRIPTION

Growth Form: The fronds arise from tough, perennial, creeping, forked rhizomes forming extensive clones.
Fronds: Overall length ± 90 cm, width ± 60 cm. Stalks ½ length of frond, grooved. Blade reflexed from top of stalk, forming a broad triangle, divided into three. The lowest pair of pinnae are so large that each comprises almost one third of the blade. Yellow green, often leathery.
Pinnae: Basal pair, subopposite, triangular, pinnules subdivided. Upper pinnae almost oblong, narrowed to a sharp point. Divisions becoming less pronounced towards the apex of blade and lower pinnae. Terminal pinnae undivided.
Pinnules: Oblong with blunt tips, often lobed at base. Sessile.
Sporangia: In sori formed along edge of pinnules, covered by a thin indusium and hidden by the rolled under leaf edge.
In most years a very small percentage of fronds produce spores.
Season: Fiddleheads appear in spring. Fronds shrivel at or before the first frost.

Table 1: The Rare Ferns of Grey and Bruce Counties

Fern		
Hart's Tongue Fern *Asplenium scolopendrium*		S3
Purple Stemmed Cliffbrake *Pellaea atropurpurea*		S3
Robert's Fern *Gymnocarpium robertianum*	S2	
Wall Rue *Asplenium ruta-muraria*	S2	
Also reported from Bruce County: **Broad Beech Fern** *Thelypteris hexagonoptera*		S3 and VUL

Note:
All the above species are considered to be nationally rare (Argus and Pryer, 1990)
Hart's Tongue Fern is rare in North America. (Flora of North America, 1990)

S2 **Very rare in Ontario**. Usually between 5 and 20 recorded occurrences in the province or with many individuals in fewer occurrences; often susceptible to extirpation.

S3 **Rare to Uncommon in Ontario**. Usually between 20 and 100 occurrences in the province or may have fewer occurrences, but with a large number of individuals in each population; may be susceptible to large scale disturbances.

VUL **Vulnerable in Ontario**. Status assigned by the Ontario Ministry of Natural Resources. Any native species that, on the basis of the best available scientific evidence, is a species of special concern in Ontario but is not a threatened or endangered species.

S ranks are internationally recognized Provincial (or subnational) ranks used by the Natural Heritage Information Centre of the Ontario Ministry of Natural Resources to set protection priorities for rare species and natural communities. The ranks are not legal designations. The NHIC evaluates provincial ranks on a continual basis and produces updated lists. Information which will assist in assigning accurate provincial ranks is always welcome.

1. Moonwort (late spring)
Botrychium lunaria

2. Mingan Moonwort (early summer)
Botrychium minganense

3. Least Grape Fern (spring)
Botrychium simplex

4. Daisy Leaf Grape Fern (spring)
Botrychium matricariifolium

5. Cut leaved Grape Fern (fall)
Botrychium dissectum

6. Blunt Lobed Grape Fern (fall) *Botrychium oneidense*

7. Leathery Grape Fern (fall)
Botrychium multifidum

8. Rattlesnake Fern (early summer)
Botrychium virginianum

50

9. Adder's Tongue (spring)
Ophioglossum pusillum

10. Cinnamon Fern (spring)
Osmunda cinnamomea

11. Interrupted Fern
Osmunda claytoniana

12. Royal Fern (early summer)
Osmunda regalis

13. Maidenhair Fern (early summer) *Adiantum pedatum*

14. Slender Cliffbrake *Cryptogramma stelleri*

15. Purple Stemmed Cliffbrake
Pellaea atropurpurea

16. Smooth Cliffbrake
Pellaea glabella

52

17. Hay Scented Fern *Dennstaedtia punctilobula*

18. Bracken *Pteridium aquilinum*

19. Northern Beech Fern
Phegopteris connectilis

20. New York Fern
Thelypteris noveboracensis

53

21. Marsh Fern
Thelypteris palustris

22. Virginia Chain Fern
Woodwardia virginica

23. Ebony Spleenwort
Asplenium platyneuron

24. Walking Fern
Asplenium rhizophyllum

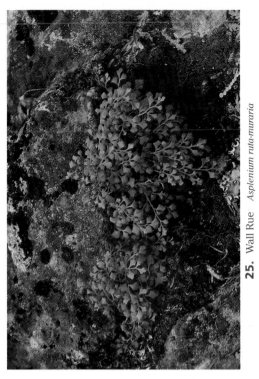

25. Wall Rue *Asplenium ruta-muraria*

26. Hart's Tongue Fern (fall) *Asplenium scolopendrium*

27. Maidenhair Spleenwort *Asplenium trichomanes*

28. Green Spleenwort *A. trichomanes-ramosum*

29. Ostrich Fern
Matteuccia struthiopteris

30. Sensitive Fern
Onoclea sensibilis

31. Narrow Leaved Glade Fern
Diplazium pycnocarpon

32. Silvery Glade Fern
Deparia acrostichoides

33. Lady Fern
Athyrium filix-femina

34. Oak Fern
Gymnocarpium dryopteris

35. Robert's Fern
Gymnocarpium robertianum

36. Bulblet Fern
Cystopteris bulbifera

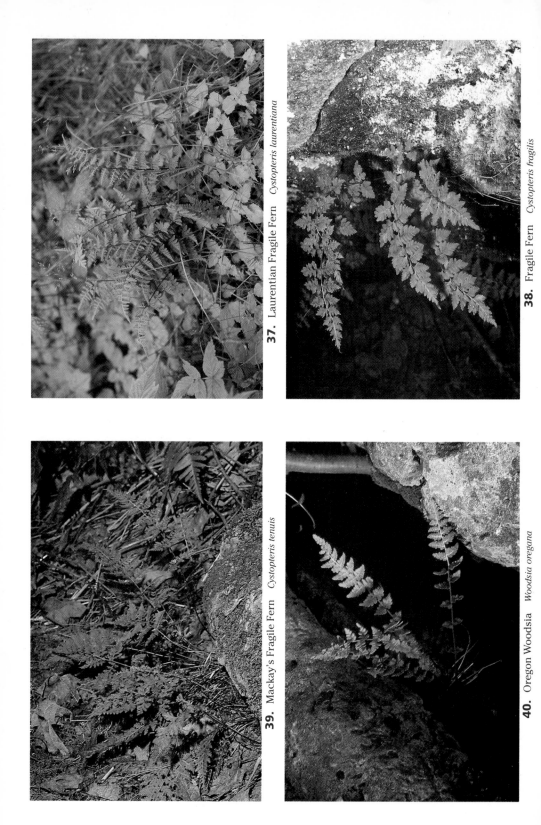

37. Laurentian Fragile Fern   *Cystopteris laurentiana*

38. Fragile Fern   *Cystopteris fragilis*

39. Mackay's Fragile Fern   *Cystopteris tenuis*

40. Oregon Woodsia   *Woodsia oregana*

41. Spinulose Wood Fern (fall) *Dryopteris carthusiana*

42. Intermediate Wood Fern *Dryopteris intermedia*

43. Crested Shield Fern
Dryopteris cristata

44. Clinton's Wood Fern
Dryopteris clintoniana

59

45. Goldie's Wood Fern (fall)
Dryopteris goldiana

46. Male Fern
Dryopteris filix-mas

47. Marginal Shield Fern
Dryopteris marginalis

48. Christmas Fern (spring)
Polystichum acrostichoides

60

49. Northern Holly Fern (spring)
Polystichum lonchitis

50. Polypody Fern (fall)
Polypodium virginianum

Historical Records Photos

(Three Fern Mysteries, see page 108)

Water Clover
Marsilea quadrifolia L.

Indian's Dream
Aspidotis densa

Curly Grass
Schizaea pusilla

MARSH FERN FAMILY
Thelypteridaceae

The Greek word *thelys* means female. Thus, *thelypteris* means **female fern**. This was the original specific name of the Marsh Fern when it was classified as *Dryopteris thelypteris*, distinguishing it from the more robust Male Fern — *Dryopteris filix-mas*. At that time, most temperate ferns were lumped together in the family **Polypodiaceae.** Taxonomists first split the genus *Thelypteris* from *Dryopteris* and then applied the name to the family when a further split was made at that level. Many older books still have the earlier arrangements. To further complicate matters, the Beech Fern has been returned to the genus *Phegopteris*, first described in 1852. In other books it may be found under *Thelypteris* or *Dryopteris* or even *Polypodium*.

Genus:
Beech Ferns
Phegopteris

In Greek *phegos* means **beech**, so literally translated this means Beech Fern. The name is derived from its European habitat where the moist, deep shade and acid soils of beech forests provide an ideal habitat for the Northern Beech Fern. Worldwide, there are only three species in this genus of which the most widespread is the Northern Beech Fern which is circumpolar. One other species is Asian but the third species, Broad Beech Fern (*Phegopteris hexagonoptera*), is a North American endemic. A substantial stand of Broad Beech Fern was reported from the Lion's Head area in 1993 by Steve Varga. Unfortunately, it could not be found in 1998. In Ontario its distribution is along the shore of Lake Erie, largely in Carolinian forest, and it is rare even there. It has been reported from as far north as Muskoka (R.O.M.). All the pinnae are winged to the axis and the frond is a much broader triangle and larger and coarser than Northern Beech Fern.

Genus:
Thelypteris

Two of our species fall into the genus **Thelypteris**. The New York Fern, *Thelypteris noveboracensis*, as its name suggests, is an eastern North American endemic. The Marsh Fern *Thelypteris palustris* is found across the northern hemisphere, but there is sufficient difference between the plants found in Europe and Asia and those found in North America for these populations to be separated as varieties or subspecies. The varietal name *pubescens* (Lawson) Fernald suggests that our specimens are rather more hairy. The typical variety is found in Eurasia, while var. *pubescens* occurs in eastern North America but farther north and further west than New York Fern. Its distribution also extends into the Caribbean. The Massachusetts Fern, though not found in our area so far, might be discovered in the future. It grows in acidic *Sphagnum* swamps so its distribution is strictly limited by habitat requirements. It is very similar in appearance to both the New York Fern and the Marsh Fern.

19. Northern or Long Beech Fern

Phegopteris connectilis (Michaux) Watt

[*Thelypteris phegopteris* (L.) Slosson, *Dryopteris phegopteris* (L.) Christens.]

Northern Beech Fern is very uncommon in Grey and Bruce Counties. It has been found in thirteen townships and then only in sparse numbers, forming small patches or clumps. It occurs in woodlands, normally swamps, but also in other lowland to upland woods, largely mixed to coniferous. Generally, it is on acid soil, sometimes near peat bogs or floating fens. Occasionally, it also grows in deciduous to conifer-dominated upland forests. It is common in the granite rock country of Muskoka where it forms large patches. This fern has yellow-green, long-stemmed, triangular fronds, crowded together and seeming to bend backwards so as to expose as much as possible of the foliage to the light. The most striking feature for identification is the bottom pair of pinnae which fork noticeably forward and downward. There is also a distinct space on the axis between them and the next pair. Above that, the pairs of pinnae merge along the axis. The small naked sori are situated along the margins of the pinnules, usually on the lower part of the frond, rarely on the rapidly tapering tip.

Northern Beech Fern, in Grey and Bruce counties, is smaller than average for this species, usually under one foot (30 cm) in length and five inches (12 cm) wide. There are usually fewer than twelve pairs of pinnae which are often hairy above and below. **Broad Beech Fern**, reported from the Lion's Head area, may be distinguished by its very broad triangular frond without the forking basal pinnules. It is much larger and coarser than Northern Beech Fern.

DESCRIPTION

Growth Form: Fronds arise from slender, branched and creeping rhizome.

Fronds: Overall length ± 20 cm, width ± 8 cm. Stalk straw-coloured with brownish scales ± ¾ length of blade. Blade triangular, tapering to a sharp point. Twice divided. Light green. Hairy.

Pinnae: Lanceolate, sharply pointed. Sessile. Upper ones merely lobed. Lowest pair characteristically pointed outward and downward.

Pinnules: Short oblong, rounded at tip. Sessile.

Sporangia: In round sori produced on veins near curled edge of underside of pinnules. No indusium.

Season: Slim, scaly fiddleheads appear in early spring through summer. Not evergreen.

20. New York Fern
Thelypteris noveboracensis (L.) Nieuwland

The New York Fern is easily identified, once enough specimens have been seen to become familiar with its key characteristics. The double tapered frond is very distinctive. Starting at the tip, the pinnae increase in length so that the widest section of the frond is near the middle; below that, the pinnae become shorter and more widely spaced with tiny, wing-like basal pinnae almost down to the ground. This gives an unmistakeable outline. Closer examination will reveal a light fuzz on the stalk. Bristle-like hairs fringe the edge of the pinnule and are also found scattered on both surfaces. Occasional hairs and glands are found on the indusium in the early part of the season. The round sori are similar to those in the genus *Dryopteris* but smaller and situated close to the margin of the pinnae at maturity. The yellow-green colour is characteristic and it is one of the first ferns to turn brown in the fall. Very often, these ferns form a thick patch as the creeping rhizomes spread and create a dense intertwining of roots with the fronds, from one to two feet (30-60 cm) high, arranged in parallel ranks facing the light.

In our area, this fern is generally found in damp woods, thickets and on the edges of swamps. Although it has a fairly wide tolerance for varying conditions, it does not grow in very wet areas and it requires a somewhat acid soil so that it is not usually found near the escarpment. One exception is in the Hope Bay area where a thick patch is growing on what is probably granitic, glacial till sandwiched between the dolostone ridges. This fern is found at various locations in Greenock Swamp but not in the wettest parts. Otherwise, it is rare in our area, especially in Grey County.

DESCRIPTION
Growth Form: Fronds arising at close intervals along a creeping, elongate, root-like rhizome. Clustered in sunny locations.
Fronds: Overall length ± 40 cm, width ± 9 cm. Stalk very short, brown and slightly scaly at base. Blade outline elliptic. Twice divided. Light yellow-green. Axis yellow green. Hairy on axis and veins. Fronds tend to face the light.
Pinnae: Lanceolate, tapering to a point. Sessile. Lowest and uppermost reduced in size.
Pinnules: Oblong, not divided to midrib. Blunt.
Sporangia: In round sori covered with tan indusium. On underside of pinnules, both sides of the vein.
Season: Fiddleheads appear in spring, fronds die off in early fall.

64

21. Marsh Fern
Thelypteris palustris Schott var. *pubescens* (Lawson) Fernald

This light green fern grows in open marshes in full sunlight with the characteristically untidy habit of having its fronds face in every direction. Finding a fertile frond is the easiest way to identify it. These appear only after the sterile fronds are well grown. The stalk is longer and the fertile pinnules seem to be pointed because the edges reflex. The lowest pair of pinnae will help distinguish Marsh Fern from New York Fern. In the Marsh Fern, the lowest pinnae are at least half the length of the largest upper pinnae, sometimes more. In the New York Fern, they are very small in comparison with the pinnae further up the blade and taper almost to the ground.

Swamps, ditches on roads going through swamps, wet meadows and streams with rich, muddy soil are typical locations, but this fern is seldom found in standing water. Under ideal habitat conditions, there is thin shade and the soil is usually slightly acidic. This medium sized fern spreads thickly from widely creeping and branching rhizomes, often forming extensive stands. New fronds are produced throughout the summer. The difference in age of the fronds, the difference between the fertile and sterile fronds and the haphazard arrangement of the blades can give the impression of several kinds of ferns growing in the one location. This is a species which can surprise us. For example, it can be found growing in a narrow dolostone crevice by the board walk on the alvar at Dyers Bay junction. Here, it forms a long hedge, growing to the exceptional height of two and a half feet (75 cm), in a habitat where you would not expect it.

DESCRIPTION

Growth Form: Fronds arising at 1-3 cm intervals from a creeping elongate, root-like rhizome.
Fronds: Overall length ± 40 cm, width ± 12 cm. Stalk straw coloured, dark at base with tan scales. ± ½ to ⅓ total length. Blade outline narrowly triangular with a few reduced pinnules above stalk. Minutely hairy. Twice divided. Fertile fronds may be even narrower and more erect with slightly reduced pinnae.
Pinnae: Long lanceolate, tapering sharply. Reduced at blade tip. Sessile.
Pinnules: Oblong, tip rounded, cut almost to midrib. Veins distinctly forked.
Sporangia: In round sori with tan indusium. On underside of pinnules, midway between vein and margin. Margins curve over when mature.
Season: Fiddleheads appear in spring, fronds die off in early autumn.

CHAIN FERN FAMILY
Blechnaceae

The ancient Greek word for ferns in general is *blechno*. This family is poorly represented in North America. Of the 10 genera and 250 species worldwide, only the genera *Blechnum* (three species) and *Woodwardia* (3 species) occur in North America, and of the six species, only two occur in temperate eastern North America.

Genus
Chain ferns
Woodwardia

This genus was named for Thomas Woodward, an eighteeenth century English botanist. The common name refers to the arrangement of the sori which are very distinct and occur in rows like chains on the mid-vein of the pinnules. Its centre of distribution is in Eurasia.

Woodwardia virginica, the name of our sole species, suggests that it was originally collected in Virginia. It is found in all the eastern coastal states as far as Texas and in the Maritimes, but it also extends around the eastern Great Lakes suggesting an oceanic climatic requirement. It is fairly close to the northwestern edge of its range in our area and so makes an exciting quest for the amateur botanist.

22. Virginia Chain Fern
Woodwardia virginica (L.) Smith

Virginia Chain Fern is habitat specific and requires an open, acidic bog where it can grow in peat moss with its roots cool and submerged. It has been found at only two stations, in southern Grey at the Wodehouse Bog and at Tobermory Bog on the Bruce Peninsula. Grey and Bruce Counties are not blessed with many acidic, boreal-type bogs. However, since we do have more than the two in which it has been found, let us start looking for this fern wherever there is an open *Sphagnum* bog.

The slightly glossy fronds are about two and one half feet (75 cm) long. This fern forms large patches. They do not grow in symmetrical crowns but often stand parallel facing the full sunlight, bearing on the reverse side the typical chain-like rows of elongated sori, parallel to the midrib. Virginia Chain Fern appears much later in spring than most ferns, another reason why it may be missed. Early in the spring, only the stubble of last year's growth is to be seen, from which the new fiddleheads will emerge much later. The fronds are about seven inches (18 cm) wide and have about twenty alternating pairs of well spaced pinnae. In the early stages, they are yellow-green but by late summer develop a darker, leathery texture with a contrasting purply-black axis. The venation of the sterile pinnule is very different from that of most ferns. Parallel to the midrib, a well defined pattern of loops or links forms, from which other veins, some free, some forking, lead to the pinnule margin. Careful observation of these features makes this a relatively easy fern to identify — if you can find it!

DESCRIPTION

Growth Form: Fronds scattered along an elongate, rope-like, creeping rhizome which has a few dark scales and fibrous roots.

Fronds: Overall length ± 75 cm, width ± 20 cm. Stalk smooth, dark, shining purple-brown, with two grooves; ± ½ length of frond. Blade lanceolate, twice divided, green, hairy and glandular when young, glands persisting. Axis purple brown fading to green or straw colour towards apex. Fertile and sterile fronds similar.

Pinnae: Elongate ovate. Sessile.

Pinnules: Cut ½ way to pinna midrib. Oblong, blunt. Margin finely serrate.

Sporangia: In elongate sori which form a double chain along pinnule midvein.

Season: Fiddleheads appear in early summer, die off in October.

SPLEENWORT FAMILY
Aspleniaceae

Since the days of Dioscorides, the Greek physician of the first century A.D., these ferns were believed to be a cure for diseases of the spleen — *splen* in Greek. Wort is the Anglo-Saxon word for herb. This family comprises about 700 species but the most recent opinion is that they all belong in the following genus.

Genus:
Spleenworts
Asplenium

This genus has, at times, been split into various segregates but the hybridization between them is such that they cannot be considered truly separate. Two of our six species of *Asplenium* were previously separated: Walking Fern as *Camptosorus rhizophyllus* and Hart's Tongue Fern as *Phyllitis scolopendrium*. Both of them, with undivided leaves, are very different in appearance from the rest of our *Aspleniums,* but most modern experts believe they belong in this group. Of our species, only Walking Fern is a North American endemic but there is a closely related species in Asia. Wall Rue, Maidenhair Spleenwort and Green Spleenwort are found in Europe. Green Spleenwort is also present in Asia. Hart's Tongue Fern is another example of a fern which has populations, occurring in both Europe and North America, that are not sufficiently different to rate as separate species. The North American variety is known as var. *americanum*. Another variety, found in Japan, is closely related to var. *americanum*. Like Wall Rue, Hart's Tongue Fern is widespread in western Europe where rock outcrops, damp stone walls and shady lanes provide favourable habitat. Ebony Spleenwort has the strangest distribution. It is widespread in eastern North America, where it appears to be migrating northwards in the Upper Great Lakes area, but it also occurs in South Africa. These strange distributions are an indication of the very ancient origins of this group. All of our species, except Ebony Spleenwort, are found here mainly in moss on dolostone.

23. Ebony Spleenwort
Asplenium platyneuron (L.) B.S.P.

Ebony Spleenwort is on the northern edge of its range in our area. It has been found from east of Durham to Sauble Beach and north to the southern portion of Lindsay Township, with about a dozen stations within that area. Probably more stands could be discovered if more people looked for them. This species appears to respond well to disturbed areas, and while once considered rare in Ontario as well as in Grey and Bruce Counties, it is actually on the increase (Cody & Britton, 1989), although still rare in this area. The preferred habitat is usually dry, open woods with partial shade, where this fern is often found growing in moss or in very shallow soil over rocks. An exception is at Sauble Beach where it has been found in heavy shade on an ancient, wooded sand dune.

The name Ebony comes from the colour of the smooth, stiff, brittle stalk. The axis, which starts out brownish-green, soon turns to shining black. The fronds taper markedly, both top and bottom. A couple of erect, ladder-like fertile fronds, one foot (30 cm) tall, appear first. They are dark green with about twenty pairs of alternating pinnae. The pinnae are well spaced, and each has a toothed edge and a well developed, ear-like lobe on the upper edge next to the axis. Later in the season, elongate spore dots will form a herring-bone pattern on the back. Three or four sterile fronds emerge next but persist into the winter. They often lie flat on the ground, and tend to be shorter than the fertile fronds. Their pinnae are less distinctly toothed and the lobe is less prominent.

DESCRIPTION

Growth Form: Fronds arising in a cluster from a short, horizontal, unbranched rhizome with dark scales.
Fronds: Stalk reddish-brown, shiny with dark, hair-like scales at base, very short ± 2.5 cm. Blade shiny, linear, tapering to both ends, once cut. Axis dark brown, smooth and shining. Fertile fronds: length ± 30 cm, width ± 3 cm. Sterile fronds ± 15 cm, width ± 2 cm.
Sterile Pinnae: Ovate, blunt, with a small ear-like lobe at base pointing to apex. Margins slightly serrate. Sessile, well spaced. Alternate.
Fertile Pinnae: Oblong, with a more distinct lobe and serrate margin. Sessile, close together. Alternate.
Sporangia: In elongate sori with a silvery indusium when young. Sori on veins in two rows angled to the midrib.
Season: Sterile fronds evergreen. Fertile fronds semi-evergreen. New sterile fronds appear during the summer.

24. Walking Fern
Asplenium rhizophyllum L.
[*Camptosorus rhizophyllus* (L.) Link)]

Walking Fern is unique among eastern North American ferns, not only in the unusual shape of its glossy, blue-green fronds, but also in the habit of producing new plants at the slender tips of the fronds. The intriguing name of this plant arouses much curiosity and expectation in the uninitiated who are often disappointed to find it does not follow them out of the woods! In reality, the Walking Fern needs at least a year to take one step, just as the domestic strawberry does. This is accomplished when the tip of the long, tapering frond produces roots, often embedded in a lush carpet of moss. In our area, it has never been seen growing on the ground, on bedrock or any substrate other than dolostone boulders; the bigger the rock, the better. Plenty of cool shade is essential too. Under the right conditions, plants will spread lavishly across the top and down the north side of huge boulders, often in a wooded ravine.

This leathery-textured frond is evergreen, lasting undamaged until well into the following season. The sori are haphazardly arranged on the underside of the frond, mostly at the base. The elongate indusia are attached to the vein and virtually disappear as maturity is reached. As befits its name, Walking Fern follows the Bruce Trail from Niagara almost to Tobermory with a last outpost on Manitoulin Island. It is uncommon on the Bruce Peninsula and rare north of Hope Bay. The Owen Sound area probably has the greatest concentration of stations in Ontario and is the best place to be sure of seeing it. In our area it often occurs many miles from the escarpment; nevertheless, it is very rare in Bruce County away from the peninsula.

DESCRIPTION

Growth Form: Fronds arising in a cluster from a small, erect, unbranched, rhizome with dark brown scales.

Fronds: Overall length ± 22 cm, width ± 2 cm. Stalk reddish-brown at base, shading to green near blade, ⅓ to ½ length of frond. Blade leathery, narrow, elongate, tapering to a sharp point, base auriculate, entire. Drooping or horizontal, tips may root. Axis dull green. Fertile leaves somewhat larger than sterile.

Sporangia: In elongate sori on the veins. Scattered irregularly, sometimes fusing.

Season: Evergreen.

25. Wall Rue

Asplenium ruta-muraria L.

Wall Rue is one of our biggest attractions, drawing fern enthusiasts to the Bruce Peninsula where it can most easily be seen in the Bruce Peninsula National Park or on Flowerpot Island. It has not been found south of Hope Bay. This tiny, delicate, deep green fern has characteristically branched, clustered fronds with small fan-shaped pinnules and toothed edges. This is our smallest fern and grows in the tiniest cracks or seams of limestone or on shallow ledges where it receives full sunlight for part of the day. It withers during a drought but recovers quickly with the first rain. It is evergreen and tends to take on a bronze shade during the winter. The spore dots are elongate and delicate with irregular edges, usually forming a sparse, random pattern, but under favourable conditions, the undersides of the leaflets may be completely covered.

The word **rue** or in Latin *ruta* was originally applied to Mediterranean shrubs used for medicinal purposes. In the British Isles, Wall Rue was considered a specific remedy for scrofula many years ago. This is a form of tuberculosis characterized by festering abscesses and swelling of the glands in the neck. This species grows prolifically on stone walls throughout Europe, **muraria** being Latin for wall. No significant difference has been established so far between the European and North American plants. It is rare in eastern North America where it is found mostly along the Appalachians, and almost always on limestone or dolostone. The stations on the Bruce Peninsula, Manitoulin Island and the Upper Peninsula of Michigan are completely isolated from the main area of distribution on this continent. It is rare here, recorded from only about fifteen locations on the upper Bruce Peninsula, almost all of them on the escarpment and always on dolostone.

DESCRIPTION

Growth Form: Fronds arising in a cluster among persistent stalk bases, from a tiny rhizome embedded in the rock.

Fronds: Overall length ± 6 cm, width ± 3 cm. Stalk reddish-brown with dark brown scales at base, shading to green. Blade outline ovate, leathery, green, thrice cut. Axis dull green.

Pinnae: Stalked, alternate.

Pinnules: Stalked, alternate. Divided into stalked, wedge shaped lobes, margins strongly toothed.

Sporangia: In oblong sori on either side of lobes, parallel to margin, ± 2 per lobe.

Season: Evergreen.

26. Hart's Tongue Fern

Asplenium scolopendrium L. var. *americanum* (Fernald) Kartesz. & Ghandi

[*Phyllitis scolopendrium* (L.) Newman var. *americanum* Fernald)]

Hart's Tongue Fern was named in England, where the typical variety occurs, from the resemblance of the strap shaped fronds to the tongue of the male Red Deer. In Ontario, this fern is considered provincially rare, but it is plentiful in Grey County and is found in some parts of the Bruce Peninsula as far north as the edge of Lindsay Township. Hart's Tongue Fern is very precise in its habitat requirements. It is found only on moist dolostone, often on bedrock in the deep shade of upland deciduous forest or in rocky ravines, sometimes in the proximity of streams or waterfalls. The dark green, leathery, elongate frond grows up to two feet (60 cm) long and two inches (5 cm) wide, with deeply cut, ear shaped lobes at the base. The sori are linear, running along the veins, forming pairs angled to the leaf axis. They sometimes approach one inch (2.5 cm) in length and are well spaced out on the back of the frond.

Newcomers to this area are greatly excited when they see this fern for the first time, especially when it is in company with the Northern Holly Fern, both growing in the same type of damp karst dolostone environment. (To a true fern fancier this is enough stimulation for one day!) In October, 1909, H.E. Ransier, botanist, photographer and writer, came from New York State to Owen Sound on the train. He arrived at a downtown hotel, paid for a room, and without delay, hired a horse and buggy for a week. He then took trips to the West Rocks, Inglis Falls, Woodford and Kemble Mountain simply to view and photograph the Hart's Tongue and Northern Holly Ferns. His article was published in the American Fern Journal under the title "Hunting the Hart's Tongue and Holly Fern in Owen Sound, Ontario, Canada." Amazingly, this trip could still be undertaken today with the same results. Hart's Tongue Fern can be seen, occurring naturally, in no other Canadian province and in only four of the United States. Ironically, one of them is New York! In North America at the present time, it is an endangered species everywhere outside of Ontario.

DESCRIPTION:

Growth Form: Fronds arising in a cluster from a short, erect, surface rhizome which is anchored by a mass of wiry roots.

Fronds: Overall length ± 30 cm, width ± 4 cm. Stalk brown to straw coloured with thick brown hairs; ± 1/5 length of blade. Blade oblong with apex rounded to a point and base auriculate; leathery, dark green, glossy. Entire. Axis brown at base shading to green colour.

Sporangia: In elongate sori, length varies. Parallel to veins on either side of the axis.

Season: Evergreen.

27. Maidenhair Spleenwort
Asplenium trichomanes L.

Maidenhair Spleenwort is one of our daintiest and prettiest ferns, as well as one of the smallest. It is often found in company with Walking Fern, growing out of crevices in moist, moss-covered bedrock or on the large, erratic, dolostone boulders (left behind by the retreating glaciers) which are often found in cool, moist woodlots. It is generally common, but very rare in Bruce County off the peninsula. The slender fronds, rarely exceeding five inches (13 cm) in length and little more than half an inch (1.25 cm) wide, are grouped in dense, spreading rosettes. There may be as many as thirty plants on one rock — a sight to behold! The stems are thread-like, fairly stiff and shining, purplish-brown in colour. The tiny pinnae, hardly more than a quarter of an inch (0.6 cm) across, are roundish or slightly oblong. They may have entire margins or slightly rounded teeth but they are not as obviously toothed as the pinnae of Green Spleenwort. The spore dots are a narrow oblong with not more than five or six to be found on the underside of each pinna.

The whole plant appears much too tender and delicate to survive the winter rigours of Grey and Bruce Counties, but the fronds overwinter and last well into the following spring. Even when the pinnae drop off, the dark, naked stems remain standing. This is clearly a rock-loving plant and will be found in our two counties wherever there is dolostone, most often avoiding sunlight.

DESCRIPTION

Growth Form: Fronds fanning out in a dense cluster from a short, partly exposed rhizome in the centre of a mass of wiry roots. Persistent leaf bases.

Fronds: Overall length ± 10 cm. width ± 1 cm. Stalks very short ± 2 cm dark purplish-brown, wiry, may be scaly at base. Blade linear, once divided. Axis purplish-brown. Sterile fronds prostrate, dark green. Fertile fronds erect, bright green.

Pinnae One side flat, rounded on side towards apex, margins slightly toothed. Almost sessile, opposite.

Sporangia: In linear sori on the veins on either side of the pinna midrib, two pairs or more making two rows at an angle with the midrib. Indusium fragile.

Season: Leaves evergreen.

28. Green Spleenwort
Asplenium trichomanes-ramosum L.
[*Asplenium viride* Hudson]

Green Spleenwort is a small, delicate fern with very specific habitat requirements, preferring cool, damp dolostone crevices or shaded, calcareous rocks on talus slopes. Often, it is associated with the very similar Maidenhair Spleenwort. The Green Spleenwort is often considered to be rare by many fern authorities. However, in Grey and Bruce Counties, there are a number of stations along the escarpment. Feversham Gorge, Bayview Escarpment, Inglis Falls, and Flowerpot Island are a few of the most notable. It may be found in other places such as the Dyers Bay Junction alvar and the Rocky Saugeen River but is much less common than Maidenhair Spleenwort.

While similar to Maidenhair Spleenwort in size and overall shape, the delicate and flexible green stalk and axis is an obvious difference. The fronds of Green Spleenwort are a little shorter and the pinnae have a more consistently deeply cut edge. Sori are only three or four per pinna. The habitat, too, is different in that Green Spleenwort requires noticeably cooler temperatures and is found deeper in narrow, shaded crevices that hold snow at the bottom until May. In these crevices there is high humidity and plenty of condensation. The less accessible habitat often causes it to be overlooked or assumed to be Maidenhair Spleenwort. This is definitely a species that requires you to go down on your knees to be sure of its identification! This fern remains quite green until the new fiddleheads appear in the spring.

DESCRIPTION

Growth Form: Fronds in a tuft, arising from a short, erect rhizome.

Fronds: Overall length ± 8 cm, width ± 1 cm. Stalk green, dark below ± 2 cm, with reddish-brown to black scales. Blade linear-oblong, once divided. Axis green. Fertile and sterile leaves similar, delicate.

Pinnae: Rhomboidal, rounded with toothed margin. Often short-stalked, subopposite.

Sporangia: In elongate sori on veins of pinnae on either side of midrib, ± 2 pairs. Fragile indusium attached on one side.

Season: Evergreen.

WOOD FERN FAMILY
Dryopteridaceae

In Greek the word *drys* means tree, the same root as in Dryad or Wood Nymph, an appropriate name, especially for the genus *Dryopteris* as many species are found on the woodland floor. This is a large family and comprises many familiar species. There are 60 genera world-wide and probably more than 3000 species. Additionally many hybrids occur when species cross. In this area we have representatives of ten genera.

Genus: *Matteuccia*

This genus was named after an Italian physicist at the University of Florence in 1866. There are only three species which are confined to temperate regions in the northern hemisphere and only one of them occurs in North America. Ostrich Fern (*Matteuccia struthiopteris*) is most prevalent in eastern North America, north of the southern limit of the Wisconsin glaciation. Its range extends across Canada to British Columbia. It is prized in many places for its edible fiddleheads and has been designated the "state vegetable" of Wisconsin! However, it is very important that these ferns not be overharvested, and they should not be eaten in large quantities because of the possibility of stomach cancer.

Genus: **Onoclea**

In Greek *onos* means container and *cleistos* means closed. This is in reference to the fertile pinnae, in which the rolled-under leaf margin encloses the sori. *Onoclea sensibilis* is the only species in this genus. It is found throughout eastern North America, east of the Great Plains, from the Boreal region to the subtropics. It also occurs in east Asia.

Genera: *Diplazium, Deparia, Athyrium*

The Greek word *diplazein* means double and refers to the double sorus which is characteristic of many species in this group but is not always present in our representative. The midrib of the pinnae is also typically deeply grooved, with the grooves running into that of the blade axis, not separated as in *Deparia*. There are 400 species in the world but only three in North America, of which our representative — Narrow Leaved Glade Fern — used to be included in *Athyrium*.

The Greek word *depas* means saucer and refers to the saucer-like indusium found in the first species to be placed in this genus. Unfortunately, this does not apply to our sole representative which has elongate sori. This mostly tropical group has only two species in North America, of which we have the Silvery Glade Fern, an eastern North American endemic. The older common name of this fern, was Silvery Spleenwort. It was also, until recently, included in the genus *Athyrium*.

The name *Athyrium* is derived from the Greek word *athyros* meaning doorless and refers to the unusual feature of this group in which the sporangia remain enclosed in the indusium until a very late stage of development. Another characteristic is the continuous grooving on the upper surface of the axis and midribs, running right into the midvein of the subleaflets. There are about 180 species of *Athyrium* worldwide but only Lady Fern remains in this genus in our area.

The genera of the Dryopteridaceae are continued on pages 81, 90, 93 and 99.

29. Ostrich Fern
Matteuccia struthiopteris (L.) Todaro

Ostrich Fern is common throughout most of Grey and Bruce Counties. This is a magnificent fern with the fronds often reaching over five feet (150 cm) in length, especially when found in its favourite habitats of low, wet woods or the banks of streams and rivers. Its fronds are distinctly dimorphic which means they have very different forms. The sterile fronds can be up to one foot (30 cm) wide with a plumelike shape (from which comes its name), broadest above the middle, distinctly rounded at the apex with a short point, and tapering gradually to the base; the stipe is very short, almost lacking. The fronds rise in symmetrical crowns from projecting rhizomes. From these, underground runners are thrown out from which spring new plants, a device which enables this plant to develop the widespread colonies which are so often found in this area. The fiddleheads from the new plants are a tasty treat in early spring when they are not more than four or five inches (10-13 cm) high and still tightly coiled. Great care should be taken not to overpick and so destroy the plants.

The fertile fronds are only two feet long, short compared to the sterile fronds. They arise in midsummer and are a bronzy-green at first but become tough, woody and brown in the fall. The pinnae are tightly rolled to protect the sori, and the spores are released in the following spring. It has often been observed that some colonies will require hard searching to locate even one fertile frond, while in others at least half the crowns will have least one in the centre. Like the related Sensitive Fern, the fertile fronds are much easier to see in winter, as they stand in contrast to the snow, and they do add an interesting touch to a winter bouquet. Since Ostrich Fern is so common, it will do no harm to pick a few.

DESCRIPTION
Growth Form: Fronds arising in a dense, circular cluster from a stout, erect branch of a scaly, dark brown rhizome.
Sterile Fronds: Overall length ± 100 cm, width ± 30 cm. Stalk stiff, brown, deeply grooved on front, scales orange-brown. Blade outline oblong, abruptly rounded at apex with a short point. Twice divided. Axis grooved on upper surface, green.
Pinnae: Lanceolate. Sessile. Subopposite.
Pinnules: Not cut to midrib of pinna, sometimes less than halfway. Oblong, blunt.
Fertile Fronds: Stalk stiff, brown, scaly at base, ± equal blade length. Blade elongate, dark green at first, becoming brown. Divided into narrow leaflets.
Sporangia: In sori on margins of fertile pinnae which roll over and form a pod-like protection.
Season: Large fiddleheads of sterile fronds appear as frost leaves the ground, leaves die off early in fall. Fertile fronds appear in early summer and persist through winter.

30. Sensitive Fern
Bead Fern
Onoclea sensibilis L.

The Sensitive Fern is perhaps the most plentiful species in wetlands in Grey and Bruce Counties as well as one of the easiest to identify. It is very common in moist, shady places but is also found in partially sunny areas. This fern is not particularly "sensitive" except to frost. Even a light frost in early autumn will cause the fronds to blacken. The coarse, broad, almost triangular fronds seem unfernlike. The pinnae protrude solidly, like the fingers of a hand, but when backlit by the sun, the prominent network of veins hidden within is revealed. The yellow stalk is usually longer than the blade, becoming brown and thickening at the base. There is a groove on the face of the stalk.

The fertile frond, which is shorter than the sterile frond, appears in midsummer. The narrow pinnae look like two parallel rows of dark green beads. This effect is created by the pinna rolling over the developing spore dots. Hence the nickname "Bead Fern." In the fall, the fertile fronds turn dark brown and become very woody. They can often be seen projecting stiffly through the snow in the depths of winter, ready to disperse their spores during the following year. Under suitable conditions, Sensitive Fern spreads rampantly as the forking rhizomes creep just below the soil surface.

This fern is particularly abundant in the southern half of the Bruce Peninsula (Albemarle, Amabel and Keppel townships) and in Greenock Swamp. In those two areas, it is clearly the most abundant ground cover plant in deciduous swamp habitat. Swamps and damp borders of streams are its favourite haunts, but patches of it occur in the most unlikely places such as dry woodlots.

DESCRIPTION

Growth Form: Fronds arising along a slender, creeping, forked rhizome, close to soil surface.
Sterile Fronds: Stalks brown at thickened base, yellow, lightly gooved above, few scales, ± 45 cm. Overall length ± 60 cm, width ± 25 cm. Blade outline broad triangle, arching back. Light yellowish-green. Once divided. Axis winged, smooth, yellow.
Pinnae: Long ovate. Sessile, joined to axis wing at base. Subopposite. Margins broadly wavy.
Fertile Fronds: Shorter than sterile fronds ± 40 cm. Stalk longer than blade. Twice divided into reduced leaflets. Green, becomes dark brown when mature.
Sporangia: In sori enclosed in rolled up pinnae — "beadlike" along the pinna.
Season: Small, red fiddleheads appear in large colonies in early spring. Fertile fronds appear in early summer. Sterile fronds die at first frost. Fertile fronds persist through winter.

31. Narrow Leaved Glade Fern
(Narrow Leaved Spleenwort)
Diplazium pycnocarpon **(Sprengel) M. Broun**
[*Athyrium pycnocarpon* (Sprengel) Tidestrom]

The Narrow Leaved Glade Fern is rare in Bruce County, although recorded as far north as Hope Bay. It is found throughout Grey, often in special "hot spots" which are notable for the number of different fern species to be found in these isolated colonies. Many ferns seem to grow in association with certain other fern species. Thus Narrow Leaved Glade Fern frequently grows in wooded ravines or sloping woodlots among stands of Goldie's Fern. Quite often, when growing under huge mature trees where other forms of undergrowth are absent, this fern spreads into extensive patches, working out and away from a thick, creeping, but seldom branching rhizome. Such stations may be found along the Massie ski trail southeast of Owen Sound, at Old Baldy in the Beaver Valley, and at the Kinghurst Forest. It appears to favour calcareous soil.

The fronds are about three feet (90 cm) long and spring up in clusters from the rhizome, the fertile ones being taller and with narrower pinnae than those of the sterile leaves. Crowded on the underside of the pinnae are two rows of well-defined linear sori, reaching from the midrib almost to the smooth margin. By late summer, this creates an obvious chevron pattern. The sterile fronds have about twenty-five pairs of long, tapering pinnae that are very thin and frail looking, allowing the forking and reforking of the veins to be easily seen. This fern is not evergreen and its frailty and scarcity suggest a keener sensitivity to uncongenial conditions than that of many other of our native ferns.

DESCRIPTION:
Growth Form: Fronds arising in a circular cluster of 5 or 6 from one end of a short, scaly, creeping rhizome lying above a mass of wiry roots.

Fronds: Overall length ± 80 cm, width ± 23 cm. Stalk scaly and dark at base, green and smooth above, ± 30 cm. Blade outline oblong-lanceolate, narrowed at base. Once divided. Axis green, slightly hairy.

Pinnae: Linear, pointed, stalked.

Sporangia: In elongate sori, occasionally double, angled to midrib on both sides of pinna — two rows. Indusium long and narrow, arching over sporangia. Brown when ripe.

32. Silvery Glade Fern

(Silvery Spleenwort)

Deparia acrostichoides **(Swartz) M.Kato**

[*Athyrium thelypterioides* (Michaux) Desvaux]

The Silvery Glade Fern is rare on the Bruce Peninsula, uncommon in southern Bruce County but not uncommon in Grey County. This is an unobtrusive fern which may easily be overlooked when hidden among its companions, the Common (Intermediate) Wood Fern and the Lady Fern. It also often occurs with Narrow Leaved Glade Fern or Goldie's Fern. The silvery-white indusia, which are its chief beauty, form a pleasing pattern of chevrons on the underside of the fertile fronds. The fronds are thin and delicate with deeply cut pinnae, almost to the midrib. They succumb quickly to early frosts.

Individual clumps may occur, but mostly, small patches are formed, spreading from a creeping rhizome which seldom branches. The fronds are two to two and a half feet (60-75 cm) long and six to seven inches (15-18 cm) wide. The stalk is one third of the frond length. It is stout and dark at the base, has abundant, narrow scales and fine hairs which become more numerous as they progress up the frond.

This fern prefers subacid soil and usually grows in rich upland (often moist) deciduous forest, sometimes near water. There are two forms found here. The first has pinnules nearly smooth-margined and round-tipped. The other has segmented pinnules which are slightly pointed at the tip. Often Silvery Glade Fern becomes confused with Lady Fern, especially when there are no fertile fronds, sometimes even one in a patch covering several square feet being difficult to detect. This is probably owing to the deeply shaded situations it seems to favour. When in sunny exposures it produces an abundance of silver-backed fertile leaves.

DESCRIPTION:

Growth Form: Rhizome short, creeping, fronds arising in a cluster at one end.

Fronds: Overall length ± 70 cm, width ± 16 cm. Stalk red-brown at base with light brown scales, green to straw colour above, ± 23 cm. Blade outline elliptic, tapering gently to a point. Fine hairs. Twice divided.

Pinnae: Lanceolate. Sessile. Subopposite.

Pinnules: Oblong, not cut to mid rib, rounded at apex, margins wavy.

Sporangia: In linear sori along veins of pinnules. Indusium silvery. At maturity spreading to cover underside of pinnules in silver, hence the common name.

Season: Fiddleheads appear in spring, leaves die at first frost.

33. [Northern] Lady Fern
Athyrium filix-femina (L.) Mertens var. *angustum* (Willd.) G.Lawson

Filix is the Latin word for fern, so *filix-femina* means, literally, lady fern, which is the very ancient common name for this fern and probably refers to its dainty aspect. This is a widespread circumboreal species with many subspecies. Much has been written about the Lady Fern, because even in our area, it has many variations of form. An attempt is made here to describe in general terms what you are most likely to see in Grey and Bruce Counties.

The light green fronds vary in length from one to four feet (30-120 cm), growing from a vase-like crown or slightly diffuse clump. While the majority of ferns send up most of their fronds in the spring, the Lady Fern continues to release additional fiddleheads until late summer. The uncoiling fiddleheads with their covering of thin, dark, hairlike chaff which persists through to maturity are an aid in identifying this plant. A fern with so many forms and variations would be a source of continual perplexity to the amateur enthusiast were it not for its distinctive, curved sori which are comma shaped, often maturing to a horseshoe form. The Lady Fern has a very untidy look by mid-summer as it succumbs to damage by insects, leaving it ragged and discoloured. The pinnae are upward ascending, and the stalk is shorter than the blade with a prominent groove up the axis of the wide frond.

The Lady Fern is very variable. One form commonly encountered is known as *forma rubellum*. This has a distinctive wine-red stalk. Two variations in shape may be described in older books. In the Upland Lady Fern, the blade tapers to a very narrow base. The Lowland or Southern Lady Fern, which does not occur in Canada, is wider towards the base. However, these variations are too inconsistent to be classified separately. At this time, all forms and variations are referred to simply as *Athyrium filix-femina*.

DESCRIPTION

Growth Form: Fronds clustered, erect and spreading from a flat, somewhat creeping, central rhizome often partly above ground and with old leaf bases attached.
Fronds: Overall length 30-120 cm, width 12-30 cm. Stalk brittle, somewhat flattened, with dark brown hairlike scales near base, light green, ± ⅓ length of frond. Blade outline usually elliptic, sometimes triangular. Thrice divided. Axis pale, smooth, may be grooved.
Pinnae: Linear, sharp pointed, usually short near stalk and much reduced towards apex. Very short stalked. Subopposite.
Pinnules: Oblong, rounded at tip, winged to pinna midrib. Subdivided into toothed lobes.
Sporangia: In horseshoe shaped sori on both sides of midvein on underside of lobes. Indusium often toothed, hairy, attached on one side to a veinlet.
Season: Fiddleheads appear in spring. Fronds die at first frost.

WOOD FERN FAMILY
Dryopteridaceae continued

Genus:
Oak Ferns
Gymnocarpium

In Greek *gymnos* means naked and *karpos* means fruit. In this genus, there is no indusium so the groups of sporangia are naked. Both our species are found in Europe and Asia as well as North America, although Robert's fern is confined to the area from Minnesota north of the Great Lakes and northeastward into Quebec, New Brunswick and Newfoundland. Oak Fern is widespread across the temperate and boreal regions of North America. The common name originated in England where oaks grow in the damp habitat favoured by this fern.

Genus:
Bladder Ferns
Cystopteris

The Greek word *kystos* means bladder and relates to the production of bladders by some species in this group. These bulblets are produced asexually (i.e. vegetative outgrowths, without fertilization) but will germinate into new plants under suitable conditions.

Genus:
Cliff Ferns
Woodsia

This genus was named by Robert Brown in 1810 after the English botanist Joseph Woods. There are 30 species which occur mostly in northern temperate regions, often at high elevations. Surprisingly, our only representative, *Woodsia oregana* ssp. *cathcartiana*, is a subspecies of a mainly western endemic named for Oregon. This subspecies ranges across the continent from California to the Bruce Peninsula so that it is on the extreme northeast edge of its range here. A much more common species north and east of this area is *Woodsia ilvensis* (L.) R.Brown — Rusty Woodsia. This is the species often described in more popular books. It is better adapted to acid substrates and the dead leaf bases are of equal length, as the stalks are jointed. Most characteristic is the wool which covers the underside of the leaves, which is white on new leaves and turns rusty brown with maturity. This wool never occurs in *Woodsia oregana*.

The remaining genera in the **Dryopteridaceae** are listed on page 90, page 93, and page 99.

34. Oak Fern
Gymnocarpium dryopteris **(L.) Newman**

In our area, Oak Fern is often found under cedars and hemlocks or close to the edge of a tiny stream, sometimes associated with Marsh Fern. A close look in this kind of moist forest will often reveal this small triune frond which looks not unlike a miniature Bracken, but much more delicate. It grows in tight groups from superficial, branching rhizomes, carpeting the spot with its lime to bright green blades, usually less than one foot (30 cm) high. Fronds are produced throughout the summer but die back in late summer or early autumn, reappearing again in the spring just as the trees leaf out. As the old rhizomes die, new clones spring up in peripheral areas.

Oak Fern fronds are broadly triangular with the axis distinctly divided into three, thus forming three, large, triangular, stalked divisions. The upper division is slightly larger and almost equilateral. The lateral divisions (strictly speaking, the lowest pair of pinnae) are slightly longer than wide, the lower pinnules are distinctly larger than their partners, and the basal pair are distinctly separated from the central axis. In the second pair of pinnae, in most cases, the basal pinnules are close to, and hide, the axis. The smooth, brittle stalks are dark at the base and longer than the blade (five to six inches or 12-15 cm) with very few scales. The blade tilts back until almost horizontal. The small, inconspicuous, naked, spore dots are borne close to the outer margin of the pinnule lobes on the underside. Oak Fern is fairly common throughout Grey and Bruce Counties wherever there is heavy shade, rich humus and plenty of moisture. It will not be found in the driest or wettest parts of such woodlands.

DESCRIPTION

Growth Form: Leaves arising singly from a dark, slender, forking rhizome with sparse, wiry roots.
Fronds: Overall length ± 22 cm, width ± 15 cm. Stalk with dark base, few scales, yellow green above, brittle, ± 15 cm. Blade ± 9 cm, outline triangular, bright green. Ternate. Thrice divided. Not glandular. Not all fronds fertile.
Pinnae: Triangular. Lower pinnae very large, slightly longer than wide, short stalked. Upper pinnae almost sessile, much longer than wide.
Pinnules: Lanceolate, distinctly longer on lower side of basal pinnae. Sessile. Lower ones, longer, deeply divided into blunt lobes. Upper less deeply divided, tapering to apex. Basal pair on basal pinnae clearly separated from central axis.
Sporangia: In tiny, round sori near upper margin of pinnules. No indusium.
Season: Fiddleheads small, delicate, three-parted. Produced throughout summer. Fronds die off in late summer.

35. Robert's Fern

or Limestone Oak Fern
Gymnocarpium robertianum (Hoffman) Newman

Robert's Fern is at its southern limit on the Bruce Peninsula where it has been found in about a dozen locations on dolostone, mostly bedrock, usually in moderate to very light shade of mixed to coniferous woods. It has an affinity for dolostone pavement areas. This fern has always been considered a rare plant and is not easy to find. When it does take hold in a small rock crevice, it gradually fills in the crevices, as the confined rhizomes branch.

In this area, the fronds are about ten inches (25 cm) long. The stalk and blade are glandular hairy, which in the early part of the season, gives the underside of the blade a "frosted" look. The blade is divided into three as in Oak Fern, but the lowest pinnae are more elongate, isosceles triangles. In Robert's Fern, the upper section is also somewhat longer than wide, and the lowest two pairs of pinnae are stalked so that the axis can be seen clearly in the centre of the frond. In Oak Fern the axis is obscured by the upper pinnules. The spore dots are small and naked and situated near the lobe margins.

DESCRIPTION

Growth Form: Fronds arising singly from a slender, blackish, scaly rhizome.

Fronds: Overall length ± 25 cm, width ± 15 cm. Stalk glandular hairy, ± ½ length of frond. Blade outline isosceles triangle. Frosty grey-green. Glandular hairy on under surface. Ternate. Thrice divided.

Pinnae: Triangular, longer than wide. Lower 2 pairs short stalked. Upper pairs more elongate triangles, very short stalked.

Pinnules: Lanceolate, pairs subequal, lower ones longer. Subsessile. Deeply divided into oblong, blunt lobes with wavy margins. Lobing distinct to apex.

Sporangia: In tiny, round sori. No indusium. Not all fronds fertile.

Season: Fiddleheads appear in spring. Fronds die off in early fall.

36. Bulblet Fern
or Bulblet Bladder Fern
Cystopteris bulbifera (L.) Bernhardi

Bulblet Fern is one of the most plentiful ferns in this area. Its graceful, lax, pendant fronds, with tips almost touching the ground, may be found anywhere with enough moisture and shade, but most often on dolostone rocks. The largest colonies seem to form by springs that flow from under the escarpment, but beautiful spreads are also witnessed in the rich, black soil of damp woodlands. Some of the most impressive stations are at the Inglis Falls Water Works springs, at Black's Park in Owen Sound and at the Devil's Pulpit on the Bruce Peninsula.

The bulblets, when present, are an easy means of identification. They are about the size of a BB shot and situated on the underside of the frond, along the midribs of the pinnae close to the axis. These fleshy bulblets drop to the ground in August and give rise to new plants directly. Reproduction is also supplemented by spores. The very long, tapering shape of the triangular fronds, as well as the fresh, bright green colour are very easily recognized. The fronds vary in length; depending on growing conditions, they may be from ten to twenty inches (30-60 cm) long and from four to six inches (10-15 cm) wide at the base. They hold their fresh, bright green colour until August, but not being evergreen, the fronds deteriorate rapidly in September. Many people are surprised to find that the little green ferns high up on the escarpment face and growing in full sunlight are most likely to be tiny Bulblet Ferns. If not, they will be Smooth Cliffbrake. Use your binoculars!

DESCRIPTION
Growth Form: Fronds arising in a cluster and arching from a short, stout rhizome with old leaf bases attached.
Fronds: Overall length ± 45 cm, width ± 12 cm. Stalk ± ⅓ length of frond, reddish when young, later greenish. Blade outline a very elongate triangle. Twice divided. Axis yellow. Glandular hairs on blade and axis.
Pinnae: Lanceolate, sharp pointed, sessile, opposite.
Pinnules: Ovate, pointed, deeply subdivided into toothed lobes.
Sporangia: In small sori on veins of pinnules, on underside near margins, 2-3 per subleaflet. Indusium hood shaped. Early fronds lack sori, later ones produce them.
Bulblets: Form on backs of pinnules in July. Shaped like tiny pistachio nuts.
Season: Slim fiddleheads appear in spring. Fronds die with first frost.

The Fragile Ferns

There are three species known as Fragile Fern found in this area. They are now all considered to be separate, but in older records all three were lumped together as *Cystopteris fragilis*. They all have distinct characteristics and habitat preferences, but they will be hard for a beginner to distinguish. *Cystopteris fragilis* is by far the most widely distributed in North America, whereas *Cystopteris tenuis* is confined to eastern North America, their ranges overlapping in our area. *Cystopteris fragilis* is most common in the Owen Sound area. Laurentian Fragile Fern which is found in the Great Lakes region is an ancient hybrid between Fragile Fern and Bulblet Fern. It has characteristics of both parents so it is often difficult to identify. It is not common, being most prevalent in the upper part of the Bruce Peninsula. Both Fragile Fern and Mackay's Fragile Fern are also ancient hybrids, but the parent they have in common is now extinct.

It should be noted that the differences among the species of *Cystopteris* are very slight. There is also the natural variability in size and form which occurs in all plants, and the possibility of hybridization to further complicate matters. *Cystopteris bulbifera* is generally easier to identify with its bulblets and very characteristic leaf shape, but even in this case, the other species intergrade with it and it is frequently found without bulblets, especially early in the season. Altogether, this group provides a challenge for even the most dedicated botanist.

Key Diagnostic Features of Cystopteris Species

Bulblet Fern — Fronds arcing. Pinnule veins mostly ending in a sinus. Axis and indusium glandular. Bulblets present on the axis. Mostly on calcareous rock and in moist woods.

Laurentian Fern — Fronds erect. Veins going to both points and sinuses. Axis and indusium glandular. Spores larger than either parent (Bulblet and Fragile). On calcareous rock.

Fragile Fern — Fronds somewhat arcing. Veins going to tips of teeth (points). Axis and indusium smooth. Thin, veiny blades and translucent stipe. Pinnule apex clearly toothed. On moist rocks, dolomitic in this area or acidic outside Grey-Bruce.

Mackay's Fern — Fronds somewhat arcing. Veins variable. Axis and indusium smooth. Indusium very small. Texture thicker and less transparent than Fragile Fern. Pinnules sometimes slightly stalked, wedge-shaped base, narrow, oblong, well-spaced. Pinnule apex only slightly, or not at all, toothed. Habitat varies: rock, rotting wood or organic soil.

37. Laurentian Fragile Fern
or Blasdell Laurentian Bladder
Cystopteris laurentiana (Weatherby)

This uncommon species is a fertile hybrid between *C. bulbifera* and *C. fragilis*. It is easier to spot on Manitoulin Island, but it has been found growing on the islands between Manitoulin and the Bruce Peninsula, where it also occurs. As well, there are other stations further south in Keppel Township and at Bayview Escarpment in St. Vincent Township.

In general, it has a larger blade than *C. fragilis* or *C. tenuis* — averaging thirteen inches by five inches (32 x 12 cm) — widest near the middle and rounded at the base. The sterile blades tend to be shorter. The lowest pair of pinnae is often spaced rather distantly from the suprabasal pair. Misshapen, scaly bulblets may occasionally be present on the upper portion of the frond. Compared to *C. fragilis* it is an upright, vigorous plant of greater stature. The pinnule veins go to both the points and the sinuses. The indusia are minutely glandular, and the spores are larger than in any other *Cystopteris* species. It can be confused with Mackay's Fragile Fern or immature, or smaller than average, Bulblet ferns, although the latter are not erect. It is - considered to be a derived species from a cross of Bulblet Fern and Fragile Fern and therefore is found mostly where the distributions of these two ferns overlap.

DESCRIPTION:
Growth Form: Fronds clustered at the apex of a short, scaly, creeping rhizome, heavily matted with dead stalks.

Fronds: Overall length ± 32 cm, width ± 12 cm. Stalk dark at base shading to reddish straw colour, slightly shorter than blade. Blade outline ovate, sharply tapered, with glandular hairs on stalk and blade. Thrice divided.

Pinnae: Lanceolate, stalked, opposite.

Pinnules: Ovate with shallow, rounded lobes. Veins ending in lobe tips and notches.

Sporangia: In sori on underside of pinnules. Indusium hood shaped with a blunt apex. Spores are larger than in either parent species. Most leaves produce spores in summer and fall.

Bulblets: Deformed ones occasionally found.

Season: Slim fiddleheads appear in early spring. Fronds die at first frost.

38. Fragile Fern
Cystopteris fragilis (L.) Bernhardi

Fragile Fern is a northern species that is found even in the Arctic. This fern is sometimes hard to separate from its two close relations, Laurentian Bladder Fern and Mackay's Fragile Fern. Their presence in a particular location is related to their varying habitat requirements as well as their geographical ranges. Fragile Fern will be the easiest to identify as it always grows on dolostone or limestone.

It has the smallest and narrowest fronds, four to eight inches (10-20 cm) in length and one to two and a half inches (2-6 cm) wide, with pairs of opposite pinnae spaced well apart, the lower pair slightly shorter than those above. The next three pairs are about the same length, giving the impression that the frond has straight sides. Other field characteristics are the translucent, brittle stalk and the thin veiny blades with veins going to the very tips of the teeth. The axis is smooth. The ends of the pinnae look slightly rounded or blunt. The indusia are lanceolate to oval and not glandular. The apparent similarity between Fragile Fern and Bulblet Fern ends upon close examination. The fronds of *C. fragilis* are not broadest at the base, have no bulblets and are much shorter than those of fully developed *C. bulbifera*. The small, unprotected, brilliant green croziers of this plant appear long in advance of true spring. Usually, the early ones are found as tiny clusters on a sheltering rock. The name fragile does not pertain to its ability to survive cold weather. Croziers continue to uncurl throughout the summer and new, light green fronds replace the early ones which turn russet and die during warm, sunny weather.

DESCRIPTION:

Growth Form: A few fronds arising in a tuft from a short, root-like, scaly, creeping rhizome matted with fibrous roots and dead stalks.

Frond: Overall length ± 20 cm, width ± 6 cm. Stalk ± ⅓ length of frond, dark at base shading to green or straw colour, translucent. Blade outline lanceolate, tapering gently to a sharp point, bright green. No hairs. Thrice divided.

Pinnae: Oblong, rounded at tip, stalked, subopposite.

Pinnules: Usually three distinct pairs at base. Ovate, rounded, with very distinct veins mostly ending in lobe tips. Subdivided into shallow, rounded lobes.

Sporangia: In sori on underside of pinnule (2-3 only). Indusium, pointed or notched, attached at one side, soon withers leaving sporangia exposed.

Most fronds produce spores in summer to fall.

Bulblets: None.

Season: Slim fiddleheads appear in early spring. Fronds die at first frost.

39. Mackay's Fragile Fern
or Mackay's Fern
Cystopteris tenuis (Michaux) Devaux
[*Cystopteris fragilis* (L.) Bernhardi var. *mackayi* G. Lawson]

Mackay's Fragile Fern was known as variety *mackayii* of Fragile Fern for many years. It is more plentiful in southern Ontario than *C. fragilis*, but they overlap in Grey and Bruce Counties, and hybridization adds to the confusion. In south Grey, Mackay's Fern is most plentiful, largely because of habitat. It will grow in habitats similar to those of Fragile Fern such as shaded rock crevices, boulders and dolostone ledges, but less often it is found on an old stump, rotted logs or in moist glades and on banks of soil.

Mackay's Fern is longer than Fragile Fern with a stalk three to eight inches (7-20 cm) long — about two thirds the length of the blade. It is dark brown at the base with a few lance shaped scales and reddish-brown or straw coloured above. The blade may be up to ten inches (25 cm) long and three and a half to four and a half inches (9-12 cm) wide. It is widest above the base and slightly convex on the sides, as opposed to Fragile Fern in which the sides of the frond are straighter. The basal pinnae are somewhat distant from the second pair, as in all the fragile ferns. Compared to *C. fragilis*, where the pinnae are rather oblong, in Mackay's they form isosceles triangles. The sori are very small, making it difficult to see the indusium which is oval. The colour is somewhat variable, but on average, Mackay's is a darker, more opaque green than Fragile Fern.

DESCRIPTION

Growth Form: Fronds clustered at the apex of a short, scaly, creeping rhizome with old leaf stalks.
Fronds: Overall length ± 25 cm, width ± 10 cm. Stalk ⅓ length of frond, dark at base, green to straw colour above. Blade outline triangular to elliptic. No hairs. Thrice divided.
Pinnae: Angled slightly upwards from axis, lanceolate, stalked, subopposite.
Pinnules: Obovate. (Widest above the middle.) Lobes very shallow and rounded. Veins ending in lobe tips and notches.
Sporangia: In sori on underside of pinnules. Indusium ovate or hood-shaped, not toothed.
Season: Slim fiddleheads appear in early spring. Fronds die at first frosts. Most leaves produce spores in summer and fall.

40. Oregon Woodsia

Woodsia oregana D.C.Eaton spp. *cathcartiana* (B.L Robinson) Windham

Oregon Woodsia has, so far, been found at only three stations in Grey and Bruce Counties, but it is frequent on Manitoulin Island. This small western fern is four to eight inches (10-20 cm) long, and the underside of the frond is glandular, although sometimes sparsely so. The indusium is divided into narrow threadlike chains of cells arching from the round sori on the backs of the pinnules. This fern is somewhat like Rusty Woodsia but Oregon Woodsia usually has no scales or hairs on the stalk or underside of the blade, so it does not develop the characteristic "rusty" look of *W. ilvensis*; nor are the stalks jointed, so the stubble of old stalks is uneven in length. Finally, Oregon Woodsia is a plant of calcareous rather than acidic soils.

In the Jones Falls area, it is located in a more or less protected, shallow crevice high on the top of the escarpment in full sunlight. The narrow fronds have opposite, triangular pinnae which are unusually well spaced. This catches the eye. It has also been found in the Cabot Head area on the Bruce Peninsula. The sterile and fertile fronds are similar. If you wish to add this fern to your "life list," September when there is less competition from other greenery, may be the best time to go hunting.

DESCRIPTION

Growth Form: Fronds arising in a small cluster among an uneven stubble of old leaf bases. Rhizome small, erect, may be partly above ground.

Fronds: Overall length ± 13 cm, width ± 3 cm. Stalk reddish-brown at base, straw colour above, pliable, not jointed, smooth or hairy, rarely with scales, ½ length of blade. Blade outline narrowly ovate, somewhat rounded at apex. Twice divided.

Pinnae: Ovate-triangular. Short stalked. Opposite.

Pinnules: Oblong, rounded at tip. 3-4 clearly defined pairs from base of leaflet. Subdivided into rounded lobes, depth of division varies.

Sporangia: In round sori, irregularly placed on underside of lobes. Indusium of thread-like lobes, often concealed by developing sporangia.

Season: Fronds die back early in the fall leaving uneven bases of old stalks. New fronds emerge in spring.

WOOD FERN FAMILY
Dryopteridaceae (continued)

Genus:
Wood Ferns, Shield Ferns
Dryopteris

This large genus of 250 species, mostly in temperate Asia, is the type genus for this family and contains some of our more conspicuous and best known ferns, a fact which is also true in Europe. Unfortunately, these species hybridize freely, thus preventing fern specialists from becoming too complacent! Much work has been done on this group in recent years and some of the species listed here will be found under other names in older books.

The Thrice Divided Wood Ferns

These lacy ferns are commonly found in moist woods throughout eastern North America. They have always created confusion among botanists, partly because of the lack of communication in the past between Europe and North America, so that many of them were named at least twice, but also because they are extremely variable and hybridize in bewildering combinations. At one time four species were lumped together as varieties of spinulose wood fern (Cobb, 1963); at other times they have been treated separately. Recent chromosome studies have done a great deal to clarify the situation. Two of the species do not concern us in this area. *D. expansa,* Northern Wood Fern, as its name suggests, is found farther north, from Lake Superior to Labrador. *D. campyloptera,* Mountain Wood Fern, occurs on the eastern seaboard and in the Appalachians.

Of the two species that are found in Bruce and Grey Counties, *D.intermedia* is endemic to eastern North America. Two closely related species are found in the Azores and Madeira. Since *D. intermedia* is the ancestral species of a number of other species in other parts of the world, this seems to suggest that it is a relic of a much wider distribution prior to glaciation. Elsewhere in the world, it has perhaps been replaced in post-glacial times by its more virile relations. Of our two representatives, it is the more common, although in North America generally, *D. carthusiana* (a tetraploid) is more widespread and is also found in Europe and Asia. *Dryopteris carthusiana* was named by the French botanist Villars in 1786 as *Polypodium carthusianum*. The name has the same origin as the Carthusian monks who came from the area of the Masif de la Chartreuse near the French-Swiss border. Various other names — most commonly *Dryopteris spinulosa,* have been attached to it by different authors during the past two hundred years, but today it is recognized as a distinct entity under the original species name. In fact, *D.carthusiana* is a hybrid derived from *D.intermedia* and another species that has vanished. Some characteristics of *D.intermedia* are intermediate between those of *D.campyloptera* and *D. carthusiana*, hence its name. It is actually a parent species of both.

Twice divided Wood Ferns — page 93

41. Spinulose Wood Fern
Dryopteris carthusiana **(Villars) H.P Fuchs**
[*Dryopteris spinulosa* (O.F. Muell.) Watt]

Spinulose Wood Fern tends to grow in somewhat wetter situations than does Evergreen Wood Fern, such as on hummocks in swampy woodlands, in thickets and on stream banks, although the distributions of the two do overlap and both can be found growing together. *D. carthusiana* is yellow-green rather than blue green and has less divergent teeth. It also looks less lacy than *D.intermedia*. The main distinguishing characteristic is found in the lowest pair of leaflets. In Spinulose Wood Fern, the lower basal pinnule next to the stalk is markedly longer than the pinnule next to it, as well as its partner above it. Therefore, at the base of the blade, on both sides of the stalk there is a very large pinnule angled out and down. Another difference is that all the pinnae in this species tend to angle upwards slightly, rather than at right angles to the axis, as in Intermediate Wood Fern. The fronds and indusia should be almost devoid of glands.

Spinulose Wood Fern is much more widely distributed than Intermediate Fern as it occurs right across Boreal-Temperate North America as well as in Eurasia. However, in eastern North America and especially in Grey and Bruce Counties, it is not so abundant as Intermediate Wood Fern. Nevertheless, Spinulose Wood Fern is common and very widespread here.

DESCRIPTION
Growth Form: Fronds arising in a circular cluster at the top of a stout, erect rhizome.
Fronds: Overall length ± 60 cm, width ± 18 cm. Stalk with oval brown scales at least at base ± ⅓ length of frond. Blade outline ovate-lanceolate, light yellow green, smooth. Not hairy or glandular. Thrice divided.
Pinnae: Angled to apex, lanceolate, asymmetrical, sharply pointed. Slightly stalked, subopposite.
Pinnules: Lanceolate, longest at base, tapering to apex of pinna. Subopposite. The lower one of the basal pair on the basal pinna is longest. Deeply subdivided into sharp lobes with forward pointing tips.
Sporangia: In round sori, two rows between margin and midvein on underside of pinnules. Indusium not glandular.
Season: Fiddleheads appear in spring, fronds sometimes persist under snow but they are not really evergreen.

42. Intermediate Wood Fern

or Common (Evergreen) Wood Fern

Dryopteris intermedia (Muhlenberg ex Wildenow) A.Gray

[*D. spinulosa* var. *intermedia*]

Graceful Wood Fern would be a more appropriate name for this elegant plant with its long, curving, lacy fronds arising in an almost perfectly circular crown. Sometimes these ferns are found growing profusely in a straight line. Closer inspection reveals the remains of a fallen tree, usually maple, that has rotted away leaving a ripple on the forest floor which nurtures dozens of these beautiful, vase-shaped, evergreen ferns. This fern is exceedingly common and can be found almost anywhere in this area where there is rich soil and ample shade, but it is most lush in hardwood stands of dolostone areas. It is frequently found on high hummocks in swamps.

The extreme variability of size and colour may make it confusing to identify. In addition, it sometimes grows mixed with *D. carthusiana* in this area, and to make matters worse, they hybridize! Novices are probably best to recognize them, as in the Peterson Field Guide (Cobb, 1963), as Spinulose Wood Fern and not try too hard to separate them. For those who wish to be more precise, a 10x lens is a great help, enabling you to see the glands which occur at least at the junction of the pinnae and blade axis of the Intermediate Fern. Under magnification the glands look like tiny stalks with a knob on top. They also appear on the indusium. The easiest way to identify this fern is to look carefully at the lowest pair of pinnae on the frond; the lower pinnule next to the stalk tends to be smaller than the pinnule next to it. This is not true of any other member of this group. It is the most common *Dryopteris* sp. in Grey and Bruce Counties. Three hybrids of *D. intermedia* quite often found in this area are:
D. x *triploidea* (*D. carthusiana* x *D. intermedia*),
D. x *boottii* (*D. cristata* x *D. intermedia*), *D.* x *dowellii*
(*D. clintoniana* x *D. intermedia*).

DESCRIPTION

Growth Form: Fronds in a circular cluster at the top of a stout, erect rhizome.

Fronds: Overall length ± 70 cm, width ± 22 cm. Stalk with tan scales at least at base, ⅓ length of frond. Blade outline oblong, with a triangular apex. Deep bluish-green, glandular. Thrice divided.

Pinnae: Asymmetrically lanceolate, wider on lower side of midrib. At right angles to axis, slightly curved up at tip. Sessile. Opposite.

Pinnules: Lanceolate, opposite. On the lowest pair of pinnae, the pinnule next to the stalk tends to be shorter than the second pinnule. Each is subdivided half way to midrib into sharp lobes.

Sporangia: In circular sori on undersurface of pinnules parallel to the midvein. Indusium with tiny glandular hairs.

Season: Fronds persisting and remaining green through the winter. New fiddleheads appearing in spring.

WOOD FERN FAMILY
Dryopteridaceae (continued)

The Twice Divided Wood Ferns

In our area there are five species of wood fern which come into this category. They have much in common with the other species of *Dryopteris*, but they appear less lacy because the pinnules are not divided into lobes — although they may be toothed on the margin.

The relationships between them are complicated. *D. goldiana*, Goldie's Fern, and *D. marginalis*, Marginal Shield Fern, are both found only in eastern North America. They are both parental species of a variety of hybrids, some of which have become established as separate species. The much more widely distributed Male Fern *D. filix-mas* is found in Europe and Asia, as well as on both sides of North America. It crosses very readily with Marginal Shield Fern suggesting an ancient relationship between these two species (Cody & Britton, 1989). For example, several dozen hybrids have been found at Skinners Bluff between the top of the escarpment and the swamp at the base, and they are also found south of the Meaford Tank Range (Pers. comm. D. Britton). *D. cristata*, Crested Shield Fern, has an extinct ancestor in common with Spinulose Wood Fern and is found in North America as well as in Europe. *D. clintoniana*, Clinton's Wood Fern, is an Eastern North American endemic which is derived from an ancient cross between Crested Shield Fern and Goldie's Fern. This has resulted in the amazing chromosome number of 246, the sum of both its parents. Humans have to get by with 46! Clinton's Wood Fern is considered to be a well defined species, although not content with having so many chromosomes, it hybridizes with a number of other species, including *D. goldiana*. It is, consequently, one of the more difficult species to identify.

43. Crested Shield Fern

or Crested Wood Fern
Dryopteris cristata (L.) A.Gray

This fern is a favourite of amateur botanists because it is easy to distinguish, except from Clinton's Wood Fern. Crested Shield Fern is always found in rich moist soil, preferably black muck and often on decomposing stumps or logs. It can be identified by the two or three tall, fertile fronds in which the well spaced pinnae twist to face the sky, thus appearing like the steps of a ladder. The sterile fronds are shorter and arching, creating an elegant surround for the stately fertile leaves. Both the sterile and fertile pinnae have a leathery texture and are deeply embossed with branching veins. On the fertile fronds, the sori are very distinct. They are generally flatter than in related species, and the kidney shaped indusium is very obvious and easily seen. This fern is not fully evergreen as the fertile fronds become relaxed at the base and fall over, although the sterile fronds remain green through the winter. Separation from the very similar, but generally larger, Clinton's Fern may be difficult. The fronds in Clinton's Fern are *not* dimorphic. A key feature in Crested Shield Fern is the difference between the larger fertile and smaller sterile fronds. The sterile fronds are narrower than in any other *Dryopteris* found in this area. Identification may be complicated somewhat by hybridization with other species of *Dryopteris*, creating hybrids, such as Boott's Fern.

Crested Shield Fern is widespread in Grey and Bruce Counties wherever wet woods and swamps occur. Although it is often found growing below cool springs that flow out from under the limestone of the escarpment, it really prefers more acid conditions. It favours similar habitat to that of Sensitive Fern, Marsh Marigold and Skunk Cabbage with which it is often found.

DESCRIPTION
Growth Form: Fronds in a circular cluster at the top of a stout, erect rhizome.

Fronds: Overall length ± 50 cm, width ± 13 cm. Stalk with scattered tan scales, at least at base, length ± ⅓ blade, outline lanceolate, tapering quite sharply, dark green not glandular. Twice divided.

Pinnae: Elongate triangle, slightly angled to apex.

Pinnules: Oblong, blunt tips, not cut to midrib of pinna. Margin serrate.

Fertile Fronds: Narrower, taller. Pinnae horizontal, lower ones smaller.

Sporangia: In round sori between margin and midvein on underside of pinnules — two rows. Indusium smooth, kidney shaped.

Season: Sterile fronds persist through winter. Fertile fronds appear in spring and fall over in the fall.

44. Clinton's Wood Fern
Dryopteris clintoniana (D.C.Eaton) Dowell

Clinton's Wood Fern may be found in habitat similar to that favoured by Crested Shield Fern but usually somewhat drier and more likely to be near limestone rock. A good example of this is in Black's Park in Owen Sound — a moist woodland area in the shelter of the escarpment. In general, this fern tends to occur in deciduous forests, on richer soil and possibly at slightly less acidic sites.

The leaves are generally few in number; often there will be only one or two fronds. They are larger and broader than in Crested Shield Fern, although the general appearance is very similar. In this species there is little difference between sterile and fertile fronds. There is a great deal of variability in size, complicating accurate identification and increasing the challenge for the dedicated fern hunter. Clinton's Wood Fern also hybridizes freely with a number of other ferns, including Goldie's Fern and Intermediate Wood Fern. In Ontario it occurs only around the lower Great Lakes and along the St. Lawrence. It is also present in Quebec and New Brunswick and a few northeastern States. It is on the edge of its range in our area.

DESCRIPTION
Growth Form: Fronds in a circular cluster at top of stout, erect rhizome.
Frond: Overall length ± 72 cm, width ± 20 cm. Stalk with dark, scattered scales at base, ± ½ length of frond. Blade outline ovate, tapering abruptly to apex, green. Twice divided. Axis with scales but no glands. Leaves variable in size. Sterile and fertile fronds similar.
Pinnae: Oblong-lanceolate, at right angles to axis, mostly straight. Subopposite.
Pinnules: Oblong with round to slightly pointed tip. Margins wavy to bluntly serrate. Not cut quite to midrib of pinna, joined to next pinnule with a narrow wing. Opposite.
Sporangia: In circular sori on either side of midvein on lower side of pinnule, two rows. Indusium not glandular, kidney shaped.
Season: Fiddleheads appear in spring. Not considered evergreen, but in this area sterile fronds often persist until spring.

95

45. Goldie's Wood Fern
Dryopteris goldiana (Hooker) A.Gray

Goldie's Fern is hunted by some enthusiasts because the thrill of finding it is akin to catching a big trout in a small stream! Eaton in 1879 described it as "one of the very finest and largest of the species" (Cody & Britton, 1989). The eyecatching fronds can be up to fourteen inches (35 cm) wide and the shape is unusual, the sides being almost parallel with a sudden narrowing to an abruptly pointed tip which arches downwards. The fiddlehead and stalk of this fern are the shaggiest of all our Northeastern American species with elongated, varicoloured scales ranging from golden-brown to blackish-striped. One of the most interesting characteristics is the colouring of the pinnae which can have golden-yellow at the midrib, gradually blending to dark green on the margin. The sori are conspicuous on the underside of the pinnules, large and evenly spaced in two parallel rows along the midvein, light green in colour changing to grey, and finally dark brown when mature.

Goldie's Fern is somewhat of a calciphile and can be found throughout the Bruce Peninsula and much of Grey County in rich, hardwood forests where there is plenty of humus and rotting leaf mulch. It is most striking when found camouflaged by more common wood ferns. In these situations a determined hunt will often reveal the elusive Narrow Leaved Glade Fern and Silvery Glade Fern growing nearby.

DESCRIPTION
Growth Form: Fronds crowded at the top of a stout, erect rhizome.
Fronds: Overall length ± 88 cm, width ± 30 cm. Stalk with dark brown to black scales, slightly less than ½ length of frond. Blade outline, broad oblong, slightly narrower at base and abruptly triangular at apex. Leathery, dark, brownish-green. Twice divided. Axis slightly scaly.
Pinnae: Lanceolate, slightly arching from axis, tips curved up. Sessile. Opposite.
Pinnules: Not quite divided to midrib of pinna, oblong, pointed. Margin serrate.
Sporangia: In round sori near midvein on underside of pinnule in two rows. Indusium kidney shaped.
Season: Fiddleheads appear in spring. Fronds die off in fall.

46. Male Fern
Dryopteris filix-mas (L.) Schott

Grey and Bruce Counties are one of the few areas in eastern North America where Male Fern is found in any quantity. It is common generally in the vicinity of the Niagara escarpment from Owen Sound north to Hopeness, although absent in some areas along this stretch. Large patches may be seen in Owen Sound, at Kemble Mountain, on Cape Croker and at Hope Bay. At Lions Head, above McKays Harbour, one thousand plants have been seen! It is also found on the Bruce Peninsula well away from the escarpment, as well as southeastwards through Grey County. It almost always occurs on dolostone.

Dryopteris filix-mas is a robust plant which may be found sprouting out of shallow crevices in open sunlight or in deep shade. The coarse fronds can be over one yard (1 metre) in length with a short, chaffy stalk. The sori are large and round, found mostly on the upper half of the frond. Characteristically, they are denser around the perimeter and thin out towards the axis. The fronds die down in December and break off, leaving the dried stalks to mark the spot where the fiddleheads will be large and well formed by the spring. This plant, under ideal conditions, has the unusual ability to multiply from the crown by creating additional, large fiddleheads crowded on top of one another, out and away from the centre, creating a very large, dense clump. The rhizomes have been used in the past to produce medicines, including a treatment for intestinal worms.

Although this is a circumpolar species, it is not widely distributed in Canada, at one time being listed as a rare species in Ontario (Argus & White, 1977). While the major part of the Ontario population is in Bruce and Grey Counties, it can also be found in adjacent Simcoe County and Michipicoten Island in Lake Superior. It would be necessary to take a trip to the Atlantic Provinces or British Columbia in order to find this fern elsewhere in Canada. It also occurs in some States, especially in the West.

DESCRIPTION
Growth Form: Fronds arising in a cluster at the top of a short, thick, scaly rhizome with dead leaf base.
Fronds: Overall length ± 80 cm, width ± 20 cm. Stalk flat on one side, green with dense dark scales, ¼ length of frond. Blade outline ovate-lanceolate, slightly tapered at the base and tapering to a sharp apex in the upper half, yellow-green. Twice divided. Axis green, flat above, scaly below.
Pinnae: Linear-lanceolate, at right angles to axis. Sessile. Opposite.
Pinnules: Not divided to midrib of pinna. Oblong, rounded tip. Margin wavy or with a few small teeth.
Sporangia: In round sori midway between margin and midvein on underside of pinnule in two rows. Indusium kidney shaped when mature.
Season: Fiddleheads appear in spring. The fronds remain until December; the dried stalks persist until spring.

97

47. Marginal Shield Fern
[Wood Fern]
Dryopteris marginalis (L.) Gray

Marginal Shield Fern is common along the escarpment and in most upland woodlots and forests in Grey and Bruce Counties. Ravines, rocky slopes and woodlots provide a foothold so long as there is rich soil and shade. In this area its roots are most often on the calcium rich dolostone, but elsewhere it is at home on granite rock and in more acid soils.

The spore dots are the key feature for identification. They are situated almost on the margins of the pinnules, an important characteristic from which both the scientific and English names are derived. The large fronds, two feet (60 cm) or more in length and widest at the base, are often seen perched high up in rocky areas on the escarpment, with the dead fronds from previous years drooping over the rock face. From below, this provides an unmistakable profile and instant recognition. This is one of the easiest ferns to view in winter, as the blue-green leaves last well into spring and are often found in high places, not completely covered by snow. During the late summer and early fall, the plant forms a crown of a dozen or more tightly rolled fiddleheads at the bottom of the "vase" of living fronds. This gives the Marginal Shield Fern a head start in spring with full sized, bright green fronds appearing before those of other species of *Dryopteris*. Later the typical blue-green colour develops. This plant does not form colonies like some ferns, but handsome individual plants live for many years, the accumulation of dead fronds helping to retain soil on the steep rocks which they inhabit.

DESCRIPTION

Growth Form: Fronds arising in a cluster from a stout, erect rhizome covered with golden-brown scales and persistent leaf bases. Matted roots often exposed.

Fronds: Overall length ± 65 cm, width ± 22 cm. Stalk ¼ to ⅓ length of frond, red-brown at base shading to green, many large golden-brown scales. Blade outline ovate, tapering to a sharp point, leathery, bluish green above, grayish green underneath. Twice divided. Axis slightly scaly, grooved above.

Pinnae: Lanceolate, slightly asymmetric. Subsessile. Opposite.

Pinnules: Oblong, rounded tip. Margins entire to lobed. Lower one of each pair slightly longer than upper.

Sporangia: In round sori near margin of pinnules, large, few in number. Indusium kidney shaped, prominent. Gray changing to dark brown on maturity.

Season: Evergreen. Fiddleheads form in fall, densely golden-brown, scaly.

WOOD FERN FAMILY
Dryopteridaceae (concluded)

Genus:
Holly Ferns
Polystichum

Poly is the Greek word for **many** and *stichos* means **line**. It is presumed that this word was coined in reference to the fact that in many species of this genus the sori line up in several rows parallel to the midvein on the pinna. There are 180 species world-wide, but only fifteen occur in North America. Many of them have a very restricted distribution. Christmas Fern and Northern Holly Fern are the only two species found in our area.

POLYPODY FAMILY
Polypodiaceae

This family contains 40 genera most of which are tropical or subtropical. Only one genus is represented here and we have only one species in that genus.

Genus:
Polypodys
Polypodium

Translated literally, this name means many little feet. The tiny scars on the rhizome, where the dead leaves have fallen off, may have made Linnaeus think of little footprints when he named this genus in 1753. In the northern hemisphere, there are many species of *Polypodium* which are closely inter-related and all look very similar. They will all be referred to as Polypody in their respective localities. The single species that we have is widely distributed in temperate and boreal eastern North America. In the northern boreal forest region it overlaps with an ancestor, *Polypodium sibiricum* which is mainly a Eurasian species. Its other ancestor is the newly recognized *P. appalachianum* found east of Grey and Bruce Counties.

(See Polypody page 102.)

48. Christmas Fern
Polystichum acrostichoides (Michaux) Schott

Christmas Fern, with its thick, evergreen, leathery fronds, is one of our best known species. In some ways it resembles the domesticated Boston Fern. This fern is widespread throughout Bruce and Grey Counties, tending to occur on rich non-rocky soil. It will be found on shaded slopes, along wooded stream banks and in ravines under mature beech and maple trees. Occasionally, it forms extensive colonies, but more often occurs as single clumps or in groups of two or three. One exceptionally large colony is at the West Rocks in Owen Sound, where within less than an acre, it is possible to count about two thousand clumps. Although Christmas Fern is not found as often on the Bruce Peninsula as it is farther south, a large colony is present behind the gravel pit on the airport road at Tobermory and another near Isaac Lake.

The fiddleheads of the Christmas Fern are among the first to appear in the spring. They are tightly coiled and partially concealed beneath long, silvery-white scales. They make a good photo opportunity at this stage. While most fronds emerge quickly in the spring, occasional sterile ones continue to appear throughout the summer in a good growing year. The average length of the sterile fronds is over two feet (60 cm) and they are four inches (10 cm) wide. Fertile fronds are the first to arise from the crown in the spring; they are taller and more erect with as many as twenty-six pairs of pinnae. Sori occur only on the pinnae of the top eight to ten inches of the frond. The pinnae in this section show a sharp reduction in size. Infertile pinnae are so prominently "eared" that they have the shape of a Christmas stocking turned sideways. This is not quite so noticeable in the common variant known as *forma incisum* in which the wavy-edged pinnae are coarsely toothed. The typical specimen will have smoother margins with fine bristly teeth.

DESCRIPTION
Growth Form: Fronds grow in an arching cluster from an erect, stout, scaly rhizome.
Fronds: Overall length ± 50 cm, width ± 12 cm. Stalk has brown base, green above, flattened in front, densely scaly, ¼ to ⅓ length of frond. Blade outline linear-lanceolate. Leathery, dark green. Once divided. Axis very scaly, green, grooved in front.
Pinnae: Oblong with rounded tip, earlike lobe on base of upper margin. Lower margin attached to axis with a small stalk.
Fertile Fronds: Sori form on reduced pinnae in upper half of blade.
Sporangia: In irregular sori with an umbrella-like indusium, densely covering underside of reduced pinnae. Red-brown on maturity. Spores smooth.
Season: Fiddleheads in early spring, stout, silvery-white, hairy. Spores produced from early summer to fall. Evergreen.

49. Northern Holly Fern
Polystichum lonchitis (L.) Roth

This fern, more than any other, excites visitors to this area when they see it for the first time. They find the heavy, somewhat rigid fronds which give the "holly" look, intriguing. The veins of the pinnae terminate in sharp, bristly spines, hence the name Northern Holly Fern. Its stature is impressive with fronds up to two feet (60 cm) in length, narrowing slightly to the base with very short, brown chaffy, stalks. They have a lustrous, dark green colour and grow evenly spaced from a prominent crown. Sterile and fertile fronds are similar in size and shape. The numerous round sori are covered with an orbicular indusium and are evenly spaced in two rows on the pinnae and the auricles. As maturity approaches, the indusium becomes funnel shaped and drops off.

Northern Holly Fern is our best example of an evergreen fern, as it remains erect until the snow gets over two feet deep. It looks striking in the spring when the brighter green new fronds form in the centre of the crown surrounded by the contrasting circle of old, dark green leaves. Along the escarpment from the Beaver Valley to Flowerpot Island, this fern usually grows on dolostone, only rarely on limestone. It has a preference for rocky woods, where it will often be found growing in rock cracks with sparse soil. This fern is found in most Grey County townships. In Bruce County it is most common towards the escarpment but becomes uncommon in the most northern two townships and is unknown south of the Bruce Peninsula. Most of the total Ontario population of this fern is found in this area.

A very unusual intergeneric hybrid of *Polystichum lonchitis* with *Dryopteris goldiana* has been found in Grey County close to the eastern boundary near the hamlet of Gibraltar. A total of five plants at three sites was discovered in the rich forest over dolostone typical of this area. This hybrid has been named x *Dryostichum singulare* W.H.Wagner (Wagner et al, 1992).

DESCRIPTION

Growth Form: Fronds erect, arching only at tip. Arising in a cluster from a stout, erect, sometimes exposed rhizome.
Fronds: Overall length ± 40 cm, width ± 8 cm. Stalk densely scaly, glandular. Short ± ⅛ length of frond. Blade outline linear, tapering to a point, widest at or above the middle. Leathery, dark green, shiny. Once divided.
Pinnae: Sickle shaped, curved on lower side, straight above with an ear-like lobe at base. Margins densely spinose-serrate, especially on upper, lobed side. Midrib off centre continuing into a short stalk. Opposite. Basal pairs very small. Sori produced on backs of middle and upper pairs.
Sporangia: In two rows of circular sori with a circular, umbrella like indusium. Sporangia spread out from under indusium and merge, coating the underside of the pinna, dark brown at maturity, contrasting with the lighter indusia. Spores spiny.
Season: Evergreen. New fronds appear in spring.

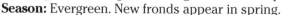

50. Polypody

[Rock Cap Fern]
Polypodium virginianum L. Polypody

As we hike along the escarpment in this area, and look down over the talus slopes, we often see Polypody softening the harsh edges of the limestone rocks. Rock Cap Fern is another, very appropriate, common name for Polypody. When conditions are ideal, its rhizomes form a dense, entangled mat across the top of large boulders where this hardy fern can withstand the dryness and exposure of such a habitat. By counting the "footprint" scars on the rhizomes, it is possible to show that these colonies, in some cases, may persist for many decades. They are a common sight where huge chunks of dolostone have fallen off the edge of the cliffs. In contrast, Polypody can also be found occasionally on large old decaying stumps and logs, and well away from the escarpment. In one case, it has been spotted forming a large patch in the woods on a steep, ancient wooded sand dune at Sauble Beach.

This fern is definitely evergreen. As winter approaches and temperatures drop, the pinnae curl more tightly, exposing the back of the frond with the mature sori. They flatten again when the warmer weather returns, but the old fronds are somewhat darker than the new ones which appear in the late spring. These old fronds remain well into the summer.

The rounded pinnae or segments are joined at the base and their edges are entire. They are nearly opposite and almost equal in size for three quarters of the frond, reducing rapidly to a blunt tip. The large round sori with no indusium form two rows, uniformly spaced, midway between the midrib and margin of the pinnae. They are found on the upper three quarters of the frond.

DESCRIPTION

Growth Form: Fronds arise in rows from a creeping, rope-like, scaly rhizome, often partly exposed, anchored by a network of fibrous roots.

Fronds: Overall length ± 25 cm, width ± 5 cm. Stalk, knobbed at base, smooth, dull green, length ± 10 cm. Blade outline oblong-lanceolate, sharply pointed, arching back. Almost once divided. Axis smooth, slightly grooved on upper surface.

Pinnae (Segments): Not cut to midrib. Oblong, rounded tip, subopposite. Margins entire.

Sporangia: In two rows of large, orange-brown sori on either side of midrib. No indusium.

Season: Scaly fiddleheads appear throughout growing season. Evergreen.

102

Table 2: The Distribution of Fern Species in Grey and Bruce Counties

On these charts the square is marked if that fern has been found **at least once** in that township unit. This gives no indication of the rarity or abundance of any species. The townships were chosen originally, rather than a more systematic grid, because some data were available. There is variation in the size of the townships but only Sarawak is markedly smaller. For the purposes of this study, the boundaries of the townships adjacent to the towns and villages have been extended through them (for adminstrative purposes the city, towns and villages are, at present, separate from the townships). For this reason we refer to the divisions on the map as "township units." Larger communities presented some problems and they have been dealt with as follows:
1. Wiarton — from Highway 6 (Berford Street) east to the bay and south of William Street (traffic signal) is included in Keppel. The remainder is in Amabel.
2. Owen Sound — Mostly in Derby. The boundary of adjacent Sydenham is defined by Highway 6/10 extended north along 9th Avenue East to 32rd Street and west down to the bay. The Derby-Sarawak Township line is extended to the Pottawatomi River and along it to the bay; north of this line is in Sarawak.
3. Those municipalities on the southern county boundaries, e.g. Lucknow and Mount Forest, have only the portion north of the streets aligned with the boundary included. Similarly, where more than one township is involved, the township lines would be extended along streets closest to the dividing line.
4. Tiverton, Hanover and Durham etc. are divided along the roads aligned with the township boundaries.
5. Records from the First Nations Hunting Reserves 60A and 60B are included in St. Edmunds. The Cape Croker First Nation 27 is included in Albemarle and Saugeen First Nations 28 and 29 are included in Amabel.
6. The boundary between the Saugeen and Amabel Township units cuts through Southampton. Going west, at the west end of Chippewa Hill, it leaves the south boundary of Saugeen First Nation 29 and follows provincial highway 21 west to the Saugeen River bridge and then west along the river to its mouth.
7. First Nation 1 is comprised of some of the islands off Oliphant and Red Bay. Burke Island and those to the south are included in Amabel. The islands north of Burke Island are included in Albemarle.

It should be noted that township boundaries are in a state of flux at this time with many amalgamations taking place, but the roads bordering the original townships will remain. The only revision we have made in this respect is to change the name of Collingwood Township and Thornbury-Clarksburg to The Town of The Blue Mountains.

Most townships have a variety of habitats, but some, especially in southern Bruce, are substantially agricultural and most of the natural habitat has been destroyed or severely fragmented.

The following individuals and organizations contributed records:
Donald Britton — University of Guelph; Mac Campbell — Saugeen Field Naturalists; Joe Johnson, Malcolm Kirk, Nels Maher, Kathy Parker, Martin Parker — Owen Sound Field Naturalists; Tom Lobb — Huron Fringe Naturalists (South Bruce).
Ministry of Natural Resources ANSI reports.
National Defence, Manoeuvre Area Planning System Report, Land Force Central Area Training Centre, Meaford.
Royal Ontario Museum Herbarium (TRT).
Grey-Sauble Conservation Authority.
Survey of southern townships conducted by the Bruce-Grey Plant Committee July-August, 1998 assisted by Ian Sinclair.

The Ferns of Bruce County

	St. Edmunds	Lindsay	Eastnor	Albermarle (Cape Croker F.N. 27)	Amabel (Saugeen F.N. 28-29)	Saugeen	Arran	Bruce	Elderslie	Kincardine	Greenock	Brant	Huron	Kinloss	Culross	Carrick
Moonwort *Botrychium lunaria* **1**	■		■													
Mingan Moonwort *Botrychium minganense* **2**	■		■	■	■											
Least Grape Fern *Botrychium simplex* **3**				■	■	■							■			
Daisy Leaf Grape Fern *Botrychium matricariifolium* **4**	■			■	■	■					■					
Cut leaved Grape Fern *Botrychium dissectum* **5**				■	■	■					■					
Blunt Lobed Grape Fern *Botrychium oneidense* **6**				■		■										
Leathery Grape Fern *Botrychium multifidum* **7**	■			■	■	■	■				■					
Rattlesnake Fern *Botrychium virginianum* **8**	■	■	■	■	■	■		■			■	■	■	■		
Adder's Tongue *Ophioglossum pusillum* **9**	■			■	■	■										
Cinnamon Fern *Osmunda cinnamomea* **10**	■			■	■	■	■	■	■		■	■	■	■	■	
Interrupted Fern *Osmunda claytoniana* **11**	■			■	■	■		■			■	■				
Royal Fern *Osmunda regalis* **12**	■	■	■	■	■	■	■	■	■		■	■	■	■	■	■
Maidenhair Fern *Adiantum pedatum* **13**	■	■	■	■	■	■	■	■	■		■	■		■	■	
Slender Cliffbrake *Cryptogramma stelleri* **14**	■	■	■	■	■											
Purple Stemmed Cliffbrake *Pellaea atropurpurea* **15**	■	■	■	■												
Smooth Cliffbrake *Pellaea glabella* **16**	■	■	■	■	■											
Hay Scented Fern *Dennstaedtia punctilobula* **17**			■								■					
Bracken *Pteridium aquilinum* **18**	■	■	■	■	■	■	■	■	■	■	■	■	■	■	■	■
Northern Beech Fern *Phegopteris connectilis* **19**	■	■	■	■	■						■					
New York Fern *Thelypteris noveboracensis* **20**				■	■	■	■				■				■	■
Marsh Fern *Thelypteris palustris* **21**	■	■	■	■	■	■	■	■	■	■	■	■	■	■	■	■
Virginia Chain Fern *Woodwardia virginica* **22**	■															
Ebony Spleenwort *Asplenium platyneuron* **23**		■	■	■	■											
Walking Fern *Asplenium rhizophyllum* **24**	■		■	■	■										■	
Wall Rue *Asplenium ruta-muraria* **25**	■	■	■													
Hart's Tongue Fern *Asplenium scolopendrium* **26**		■	■	■	■						■	■			■	
Maidenhair Spleenwort *Asplenium trichomanes* **27**	■	■	■	■	■		■				■				■	■
Green Spleenwort *A. trichomanes-ramosum* **28**	■	■	■	■	■											
Ostrich Fern *Matteuccia struthiopteris* **29**	■	■	■	■	■	■	■	■	■	■	■	■	■	■	■	■
Sensitive Fern *Onoclea sensibilis* **30**	■	■	■	■	■	■	■	■	■	■	■	■	■	■	■	■
Narrow Leaved Glade Fern *Diplazium pycnocarpon* **31**			■	■	■			■						■	■	
Silvery Glade Fern *Deparia acrostichoides* **32**		■	■	■	■			■	■	■	■	■		■	■	
Lady Fern *Athyrium filix-femina* **33**	■	■	■	■	■	■	■	■	■	■	■	■	■	■	■	■

The Ferns of **Bruce County**

Fern	#	St. Edmunds	Lindsay	Eastnor	Albemarle (Cape Croker F.N. 27)	Amabel (Saugeen F.N. 28-29)	Saugeen	Arran	Bruce	Elderslie	Kincardine	Greenock	Brant	Huron	Kinloss	Culross	Carrick
Oak Fern *Gymnocarpium dryopteris*	34	■	■	■	■	■	■	■	■	■		■	■			■	■
Robert's Fern *Gymnocarpium robertianum*	35	■	■														
Bulblet Fern *Cystopteris bulbifera*	36	■	■	■	■	■	■	■	■		■	■	■	■		■	■
Laurentian Fragile Fern *Cystopteris laurentiana*	37	■	■	■	■				■								
Fragile Fern *Cystopteris fragilis*	38	■	■	■	■	■										■	
Mackay's Fragile Fern *Cystopteris tenuis*	39	■	■	■	■												
Oregon Woodsia *Woodsia oregana*	40		■														
Spinulose Wood Fern *Dryopteris carthusiana*	41	■	■	■	■	■	■	■	■	■	■	■	■	■	■	■	
Intermediate Wood Fern *Dryopteris intermedia*	42	■	■	■	■	■			■	■						■	■
Crested Shield Fern *Dryopteris cristata*	43	■	■	■	■	■	■	■	■	■	■	■	■	■	■	■	■
Clinton's Wood Fern *Dryopteris clintoniana*	44		■	■	■	■	■		■	■			■		■	■	
Goldie's Wood Fern *Dryopteris goldiana*	45		■	■	■	■	■						■			■	
Male Fern *Dryopteris filix-mas*	46	■	■	■	■	■											
Marginal Shield Fern *Dryopteris marginalis*	47	■	■	■	■	■	■	■	■	■	■	■	■	■	■	■	■
Christmas Fern *Polystichum acrostichoides*	48	■	■	■	■	■	■	■	■			■	■	■	■	■	
Northern Holly Fern *Polystichum lonchitis*	49	■	■	■	■	■											
Polypody Fern *Polypodium virginianum*	50	■	■	■	■	■	■				■	■	■				

Note:
The square is marked if the fern has been found at least once in that township unit.
Every endeavour has been made to ensure the authenticity of these records. However, earlier studies did not always distinguish clearly among *Botrychium lunaria, B. minganense* and *B. simplex* (see pages 27, 28 and 29). Similarly, *Cystopteris tenuis* was only recently separated from *C. fragilis* (see p. 87, 88).

If anyone discovers in a township unit a fern that is not recorded here, please inform the M.N.R.
(See form at the back).

The Ferns of Grey County

	Keppel	Sarawak	Derby	Sydenham	St. Vincent	Sullivan	Holland	Euphrasia	Town of Blue Mountains (Collingwood Twp.)	Bentinck	Glenelg	Artemesia	Osprey	Normanby	Egremont	Proton
Moonwort *Botrychium lunaria* 1																
Mingan Moonwort *Botrychium minganense* 2	■															
Least Grape Fern *Botrychium simplex* 3	■										■					
Daisy Leaf Grape Fern *Botrychium matricariifolium* 4	■			■			■		■		■	■	■			
Cut leaved Grape Fern *Botrychium dissectum* 5	■		■	■	■			■	■		■		■			■
Blunt Lobed Grape Fern *Botrychium oneidense* 6					■		■		■							
Leathery Grape Fern *Botrychium multifidum* 7	■		■	■			■				■		■			■
Rattlesnake Fern *Botrychium virginianum* 8	■	■	■	■	■	■	■	■	■	■	■	■	■	■	■	
Adder's Tongue *Ophioglossum pusillum* 9			■						■							
Cinnamon Fern *Osmunda cinnamomea* 10	■		■	■	■	■	■	■	■	■	■	■	■	■	■	■
Interrupted Fern *Osmunda claytoniana* 11	■		■		■	■	■		■	■	■	■	■			
Royal Fern *Osmunda regalis* 12	■		■	■	■	■	■	■	■		■		■	■	■	■
Maidenhair Fern *Adiantum pedatum* 13	■	■	■	■	■	■	■	■	■	■	■	■	■	■		
Slender Cliffbrake *Cryptogramma stelleri* 14	■	■	■	■	■			■	■		■		■			
Purple Stemmed Cliffbrake *Pellaea atropurpurea* 15																
Smooth Cliffbrake *Pellaea glabella* 16	■	■	■	■	■			■	■		■	■				
Hay Scented Fern *Dennstaedtia punctilobula* 17				■			■	■					■			
Bracken *Pteridium aquilinum* 18	■	■	■	■	■	■	■	■	■	■	■	■	■	■	■	■
Northern Beech Fern *Phegopteris connectilis* 19	■		■					■	■		■	■	■			
New York Fern *Thelypteris noveboracensis* 20	■		■	■	■											■
Marsh Fern *Thelypteris palustris* 21	■	■	■	■	■	■	■	■	■	■	■	■	■	■	■	■
Virginia Chain Fern *Woodwardia virginica* 22								■								
Ebony Spleenwort *Asplenium platyneuron* 23	■		■	■							■					
Walking Fern *Asplenium rhizophyllum* 24	■	■	■	■	■	■	■	■	■	■	■	■				
Wall Rue *Asplenium ruta-muraria* 25																
Hart's Tongue Fern *Asplenium scolopendrium* 26	■	■	■	■	■			■	■		■		■		■	
Maidenhair Spleenwort *Asplenium trichomanes* 27	■	■	■	■	■		■		■		■	■	■			
Green Spleenwort *A. trichomanes-ramosum* 28	■	■	■	■	■		■		■	■	■	■	■			
Ostrich Fern *Matteuccia struthiopteris* 29	■	■	■	■	■	■	■	■	■		■	■		■	■	■
Sensitive Fern *Onoclea sensibilis* 30	■	■	■	■	■	■	■	■	■	■	■	■	■	■	■	■
Narrow Leaved Glade Fern *Diplazium pycnocarpon* 31	■	■	■	■	■	■	■	■	■	■	■	■				
Silvery Glade Fern *Deparia acrostichoides* 32	■		■	■	■	■	■	■	■		■	■			■	
Lady Fern *Athyrium filix-femina* 33	■	■	■	■	■	■	■	■	■	■	■	■	■	■	■	■

The Ferns of Grey County

	Keppel	Sarawak	Derby	Sydenham	St. Vincent	Sullivan	Holland	Euphrasia	Town of Blue Mountains (Collingwood Twp.)	Bentinck	Glenelg	Artemesia	Osprey	Normanby	Egremont	Proton
Oak Fern *Gymnocarpium dryopteris* 34	■	■	■	■	■		■	■	■	■	■	■	■	■	■	■
Robert's Fern *Gymnocarpium robertianum* 35																
Bulblet Fern *Cystopteris bulbifera* 36	■	■	■	■	■	■	■	■	■	■	■	■	■	■	■	■
Laurentian Fragile Fern *Cystopteris laurentiana* 37	■				■											
Fragile Fern *Cystopteris fragilis* 38	■	■	■	■	■		■	■	■	■	■	■				
Mackay's Fragile Fern *Cystopteris tenuis* 39	■	■	■	■	■		■	■			■					
Oregon Woodsia *Woodsia oregana* 40			■													
Spinulose Wood Fern *Dryopteris carthusiana* 41	■	■	■	■	■	■	■	■	■	■	■	■	■	■	■	■
Intermediate Wood Fern *Dryopteris intermedia* 42	■	■	■	■	■	■	■	■	■	■	■	■	■	■	■	■
Crested Shield Fern *Dryopteris cristata* 43	■	■	■	■	■		■	■	■	■	■	■	■	■	■	■
Clinton's Wood Fern *Dryopteris clintoniana* 44	■		■	■	■		■	■	■		■					■
Goldie's Wood Fern *Dryopteris goldiana* 45	■	■	■	■	■	■	■	■	■	■	■	■	■			
Male Fern *Dryopteris filix-mas* 46	■	■	■	■	■	■	■	■			■					
Marginal Shield Fern *Dryopteris marginalis* 47	■	■	■	■	■	■	■	■	■	■	■	■	■	■	■	■
Christmas Fern *Polystichum acrostichoides* 48	■	■	■	■	■	■	■	■			■	■	■		■	
Northern Holly Fern *Polystichum lonchitis* 49	■	■	■	■	■		■	■	■			■		■		
Polypody Fern *Polypodium virginianum* 50	■	■	■	■	■		■	■	■	■	■	■	■			

Note:

The square is marked if the fern has been found at least once in that township unit.

Every endeavour has been made to ensure the authenticity of these records. However, earlier studies did not always distinguish clearly among *Botrychium lunaria, B. minganense* and *B. simplex* (see pages 27, 28 and 29). Similarly, *Cystopteris tenuis* was only recently separated from *C. fragilis* (see p. 87, 88).

If anyone discovers in a township unit a fern that is not recorded here, please inform the M.N.R.
(See form at the back).

Three Fern Mysteries

Three fern species which were recorded many years ago from this area have never been found again. The most recent was *Marsilea quadrifolia* L. or **Water Clover**, so-called because the long-stalked leaves with four leaflets look exactly like a Four Leaved Clover and are about the same size. This species has a long thin rhizome which creeps along the bottom of shallow, sluggish streams, sending up leaves at intervals. The sori are enclosed in a tiny, stalked, oval structure which looks somewhat like a flattened capsule. This arises at the base of the leaf stalk and is only about 4 mm long.

This Eurasian species was introduced into Connecticut in 1860. Since then it has spread from the New England States into New York State and southern Ontario and from there westward as far as Missouri. It was discovered in a small stream close to Highway 70 (now Highway 6) on a farm at Springmount, Derby Township, about twenty-five years ago. According to Cody and Britton (1989), it is used in artificial ponds and aquaria from which it may escape temporarily. This seems the most likely explanation for its presence in Grey County. Alternatively, the spores may have been brought in by one of the many tourist vehicles coming from farther south to visit the Bruce Peninsula. It has not been seen since, but the present trend to an increase in average temperature may make conditions more favourable so that it could crop up again. It is worth bearing this in mind and looking out for it. Anything that looks like a half-drowned clover is a possible candidate. The clues would be the four lobes and the tiny "capsules."

The oldest of the three records dates back to 1883, when Dr. Henry Ami of the Geological Survey of Canada collected *Aspidotis densa* (Brackenridge in Wilkes) Lellinger [recorded, at the time, as *Cheilanthes siliquosa* Maxon] near Durham in Grey County. This is mainly a western species which rejoices in the common name of **Indian's Dream**! It does occur also in the Gaspé Peninsula and southern Quebec, possibly an example of a once, more widely distributed species. The authenticity of this record has been questioned, although the specimen does exist and is correctly identified. The label reads "rocky hillsides of Guelph dolomite, Durham, Ontario." It has, however, never been found again. This small, rather leathery fern, at most 30 cm tall, is related to *Pellaea* and *Cryptogramma,* but the pinnae are thrice divided. The pinnule lobes are very linear and sharp tipped. The sori make a line along the margins of the lobes except at the apex.

Most intriguing, and intermediate in date, is *Schizaea pusilla* Pursh or **Curly Grass** which was reported from the Sauble Beach area on Lake Huron in Bruce County in 1928 by Eugene Moxley, an Owen Sound resident who was a keen, and very knowledgeable, amateur bryologist and a member of the Sullivant Moss Society. Accounts vary as to what actually happened, but according to the annotation with the specimen in the herbarium in Toronto (TRT), the story goes something like this. In the spring of 1928, Moxley was asked by a botanist named McColl to collect some *Selaginella selaginoides* for him. He knew that this spikemoss grew in the wet swales between the sand dunes at Sauble Beach, but when he got there, he found that the ground was still frozen so he hacked out some lumps of soil and took them back home. When he thawed them out, he found not only the *Selaginella* but also a curious little plant consisting of wiry, curled, grass-like fronds with taller, erect structures looking, in profile, rather like toothbrushes. By July 1928 it was pressed and mounted, and he had identified it as Curly Grass, but when he proudly showed it to his botanical friends, no one believed he had found it at Sauble Beach. They said he must have mixed up his specimens with some from elsewhere. It was well known to be an Atlantic coastal species found only in Nova Scotia, Newfoundland and the New Jersey Pine Barrens where it grows in sandy or peaty depressions, in habitat similar to that found at Sauble Beach.

Naturally, Moxley was very upset by the reception of his news so he packed up the specimens and put them away, refusing to discuss the matter further. Luckily, being a typical botanist, he did not throw them away. By coincidence, a management shuffle in the firm for which he worked caused him to move to Toronto very shortly thereafter, and he deliberately left his specimens behind in the attic of his house, where they remained until someone had a good clean out in 1945 and returned them to him. He promptly sent them to his friend Hubert H. Brown who believed his story implicitly and published an account in volume 35 of the Fern Journal. By then, development of the Sauble Beach area had taken place and the drainage patterns had changed. Exhaustive searching by knowledgeable botanists has failed to reveal any more specimens.

Prospective collectors commented on the similarity between Sauble Beach and the Pine Barrens. Atlantic coastal species, relicts of post-glacial lake shores and past climates, do occur not too far away in Muskoka. *Aspidotis densa* has been found in the mountains of the Gaspé Peninsula so it is possible, in both cases, that these were the last lingering remnants of much more extensive post-glacial distributions. On Hubert Brown's death Moxley's specimen, among other collections, was donated to the University of Toronto herbarium, now managed by the Royal Ontario Museum, where it may be seen to this day.

Note: A more detailed account by William G. Stewart, who was well acquainted with a close friend of Eugene Moxley, may be found in The Field Botanists of Ontario Newsletter, Volume 10 (3), Fall 1997.

(See page 61 for historical fern photos).

GROWING FERNS FROM SPORES

The time to harvest fern spores is late summer to early fall. Pick a fertile frond of the species you wish to reproduce. Colour is a reliable clue to ripeness; as the spores mature, the sori darken to a deep, rich brown. Lay the frond spore-side down on a sheet of clean white paper. Cover the frond with another sheet of paper and place a paperweight on top. Within a day or two the sporangia will open and release their spores.

Prepare a sterile nursery. To germinate successfully, spores require 100 percent humidity and sterile conditions, because they cannot compete successfully with algae, fungi, or bacteria. Obtain a transparent plastic tray with no compartments and a tightly fitting lid. Sterilize your container by scrubbing it with a 10 percent solution of chlorine bleach. Next put a deep layer of commercial peat moss in the tray. Saturate with boiling water (tap water contains algae). Cover the container and set it aside to cool. Purchase sterile potting soil which still needs three hours in the oven at 300°F in a covered container. Put two inches of the sterilized potting soil on top of the peat and add boiling water to saturate.

Sow the spores. A day or two after completing step 1, remove the paperweight, the top sheet of paper, and the drying frond. You should find the bottom sheet of paper covered with a fine dust of spores. Carefully remove the chaff of the indusium, crease the paper down the centre to form a trough into which the spores can settle. Open the cooled nursery box, tip the paper over it, and tap with your finger to gently sift the spores over the top of the soil. Spread the spores as evenly as possible covering the entire surface. Replace the lid immediately and tape down with masking tape.

Germination. The best place for the tray is a north-facing window where it will receive bright but indirect sunlight (direct sunlight will cook the spores). Fluorescent light is also fine. Germination usually begins in three or four weeks, and a film of pale green should cover the growing medium. Another month or two should see the appearance of recognizable prothalli. When these stop growing, little leaves should emerge. If they do not, it may be that conditions within the tray are too dry for the sperm to swim to the eggs. Undo the tape and try misting the prothalli with distilled or boiled and cooled water.

Transplanting. As soon as little fronds appear, prepare other sterile boxes, this time covering the peat moss layer with a pasteurized commercial African-violet potting mix. Don't try to separate the plants; prick them out in little clumps (the tip of a knife makes a good lifting tool). Replant these plugs in rows at one-inch intervals. Water as necessary, with sterile water — the growing medium should stay moist but not be soaked. When the young ferns reach a height of 2 inches and danger of frost has passed, dig them up. Separate them, and transplant them to a shady out door bed appropriate for that specific fern.

GLOSSARY

Acute Sharp pointed — as in a pinna or pinnule.

Alternate Pinnae on each side of the axis *not* opposite. Similarly, pinnules not opposite as they arise from the pinna.

Alvar Flat, unforested area of smooth limestone pavement, usually extensively creviced.

Apex Tip of leaf or leaflet.

Asexual Reproduction A form of reproduction without reduction division and fusion of cells.

Auriculate With ear-like lobes.

Axis The midrib of the fern blade.

Basal At the base of a structure; e.g. basal pinnae are at the base of the blade.

Biennial Of plants which complete their life cycle in two years.

Blade The leafy or expanded portion of the fern frond.

Bog An open, acid wetland which is poor in nutrients. Usually unforested and dominated by *Sphagnum*.

Bulblet A tiny, asexual reproductive structure produced on the fronds of some ferns.

Calcareous Of rocks or soil containing a high proportion of calcium compounds.

Calciphile A species requiring a substrate with high calcium content.

Carcinogenic Referring to substances known or suspected to cause cancer.

Circumboreal Found around the northern hemisphere in the Boreal Forest region, south of the Arctic circle and, strictly, north of the Temperate Decidous Forest.

Chaff or Chaffy With scales resembling wheat chaff on the stalk of the fern.

Clone A colony of plants arising from a single individual by asexual means.

Crozier An alternative term for "fiddlehead."

Denticulate The edge of a pinna or pinnule which has very small teeth (directed outward).

Dimorphic Two forms. Applied to ferns that have fertile and sterile fronds that look distinctly different.

Diploid Of cells which have two sets of chromosomes. Formed when the sex cells unite.

Dolostone The commonest exposed bedrock in Grey and Bruce Counties. Composed largely of dolomite (Magnesium calcium carbonate). Slightly harder and more resistant to weathering than limestone (Calcium carbonate).

Drumlinized Glacial deposits formed into ridges paralleling the advance of the ice.

Elliptic	Of a pinna or pinnule with approximately the rather long oval outline of an ellipse.
Escarpment	*See* Niagara Escarpment.
Eskers	Ridges of sand and gravel formed in the channels of melting glaciers and deposited as the ice retreated.
Fen	An open wetland with slowly moving water and low vegetation on a non-acid substrate. In some cases vegetation forms a floating mat.
Fiddlehead	Used to describe the developing fern frond in its curled up stage. (*See* also Crozier).
Frond	The fern "leaf" including the blade and stalk which arise from the rhizome.
Gametophyte	A tiny, simple plant formed by the germinating spore and containing only half the chromosomes of the parent plant, i.e. **haploid**. The sex cells are produced in special organs. After fertilization, the sporophyte will grow from the gametophyte surface. As it matures the gametophyte disappears.
Glabrous	Without hairs, smooth.
Glandular	Of blades or stems bearing glands, often in the form of hairs, which will secrete substances in liquid form. These may serve to remove substances toxic to the plant and may make the fern unpalatable to grazing animals.
Haploid	Of cells which contain only one set of chromosomes. In ferns, formed at spore development and found in the gametophyte.
Indusium	Flap of tissue which may be formed with the sorus to protect the sporangia.
Karst	A type of topography developed in limestone and dolostone areas, caused by chemical dissolution of the rocks by water. Characterized by pitting, sink holes and underground streams.
Lanceolate	The long, narrow, pointed shape of a pinna or pinnule.
Leaflets	Subdivisions of the blade, i.e. **pinnae**
Lobe	A distinct subdivision of a pinnule, making the blade thrice divided — more than merely toothed.
Margin	Edge of leaf.
Meiosis	See Reduction Division.
Midrib	The central vein of a pinna or a pinnule.
Niagara Escarpment	The geological formation which runs through southern Ontario from Niagara to Tobermory and on through Manitoulin Island. Where exposed, as it is through most of Grey and Bruce Counties, it forms a steep ridge often with cliffs on the Georgian Bay side and a gentler slope towards Lake Huron on the other side.

Oblong	Referring to pinnae or pinnules which are parallel sided, with rather blunt ends.
Obtuse	Referring to pinnae or pinnules with a blunt apex.
Once Divided	Referring to fern blades divided into simple pinnae, i.e. pinnate
Opposite	Pairs of pinnae or pinnules arising exactly opposite one another on the axis.
Orbicular	Referring to pinnae or pinnules approximately circular in outline
Ovate	Oval frond, pinna or pinnule, widest at base.
Persistent	Referring to structures such as fronds or bases of stalks which last into the next season.
Pinna	A subdivision of a fern frond (plural **pinnae**) i.e. a "leaflet."
Pinnule	A subdivision of a pinna, i.e. a "subleaflet." May be further subdivided into lobes.
Perennial	A plant which grows and reproduces for many seasons.
Rachis	The correct term for the midrib or axis of a fern blade.
Ravine	A long, narrow, forested cleft between heights often with a creek or spring run-off.
Reduction Division	A cell division in which the two sets of chromosomes separate so that the resulting cells are haploid. In ferns, this takes place in spore formation.
Reflexed	Bent backwards.
Revolute	Rolled under as in the edge of a leaf or leaf division
Rhizome	An underground stem, often horizontal and creeping, from which the roots and fronds arise. Sometimes partially above ground.
Rhomboidal	More or less the shape of a rhombus, i.e. a parallelogram with equal sides.
Rosette	A circular cluster of fronds — shaped like a rose.
Scale	A small outgrowth on the stalk, often characteristically shaped or coloured.
Serrate	Saw-toothed, i.e. the teeth directed forward (*see* denticulate).
Serrulate	With a very fine serrate edge.
Sessile	Without a stalk.
Sinus	The indent between pinnules or lobes.
Sorus	Cluster of sporangia (plural **sori**).
Spike	Used to refer to the stalk with the cluster of fertile pinnae which comprises the fertile portion of the frond in some fern species, e.g. Adder's Tongue.
Spinose	Armed with spines.
Sporangium	The tiny stalked organ in which the spores form (plural **sporangia**).

Spores	Single cells formed by **reduction division** (meiosis) from which the gametophyte may develop. Containing half the chromosomes of the parent plant.
Spore Dots	Used to refer to the circular sori on the backs of the fronds in some fern species.
Sporophyte	The most visible stage of the life cycle in ferns, containing a double set of chromosomes, i.e. diploid, and producing the spores.
Stipe	The correct term for the stalk of a fern frond.
Sub–	A prefix used to indicate "rather less than," e.g. subacute.
Subleaflets	The lobes produced by the division of the leaflets, i.e. **pinnules**. These may be toothed or further subdivided into lobes.
Subopposite	Pairs of leaflets or subleaflets not quite opposite as they arise from the axis.
Succulent	Referring to plants or parts of plants which are thick and fleshy due to the presence of water storage tissue.
Supra–	A prefix used to indicate "rather more than."
Swamp	Forested wetland.
Talus	The sloping mass of rock fragments at the base of a cliff.
Ternate	Divided into three more or less equal sections.
Tetraploid	Referring to cells which have four sets of chromosomes. Formed by doubling of the diploid number. This sometimes occurs in hybrids, e.g. the tetraploid *Dryopteris carthusiana* has a sporophyte chromosome number of 164. In one of the ancestral species, *D. intermedia,* the sporophyte chromosome number is 82.
Thrice Divided	Referring to ferns with pinnae subdivided into pinnules and the pinnules further subdivided into lobes.
Triune	A group of three which is actually one.
Twice Divided	Referring to ferns with pinnae divided into pinnules — pinnules may be toothed but not lobed.
Veins	The water conducting (vascular) strands in a frond (leaf) or its subdivisions.

REFERENCES AND BIBLIOGRAPHY

Argus, G.W. & Pryer, K.M. 1990. **Rare Vascular Plants in Canada.** Canadian Museum of Nature. Ottawa.

Argus, G.W & White, D.J. 1977. **The Rare Vascular Plants of Canada.** Syllogeus 14: 1-63. Canadian Museum of Nature, Ottawa.

Brown, Hubert, H. 1945. *Schizaea pusilla* **from Ontario, Canada.** Amer. Fern. J. 45: 40-41.

Bruce-Grey Plant Committee. 1997. **A Checklist of Vascular Plants for Bruce and Grey Counties, Ontario.** 2nd edition Owen Sound Field Naturalists.

Chapman, L.C. & Putnam, D.F. 1984. **The Physiography of Southern Ontario.** ON Geol. Surv. Ontario M.N.R.

Cobb, B. 1963. **A Field Guide to the Ferns.** Peterson Field Guide Series. Houghton Mifflin Company, Boston.

Cody, W.J. and Britton, D.M. 1989. **The Ferns and Fern Allies of Canada.** Agriculture Canada.

Flora of N. Am. Ed. Comm. 1993. **Flora of North America Volume 2 — Pteridophytes and Gymnosperms.** Oxford University Press, New York

Gillespie, J.E. & Richards, N.R. 1954. **Soil Survey of Grey County.** Report No. 17 ON Soil Survey.Canada Dept. of Agriculture & the Ontario Agricultural College

Gleason, H. et al. 1952. **New Britton & Brown Illustrated Flora.** Lancaster Press. Penn.

Hallowell, Barbara G., 1981. **Fern Finder.** Nature Study Guide. Berkeley, California.

Hodge, W.H. 1973. **Fern Foods of Japan and the Problem of Toxicity.** Am. Fern J. 63: 77-80.

Hoffman, D.W. & Richards, N.R. 1954. **Soil Survey of Bruce County.** Report No. 16 ON Soil Survey. Canada Dept. of Agriculture & the Ontario Agricultural College, Guelph.

Johnson, J. 1990. **Vascular Flora of Three Regions Comprising Bruce and Grey Counties, Ontario, With Emphasis on Rare Taxa.** Ontario Ministry of Natural Resources.

1996. **Landsat Image Classification Natural Features Mapping.** Ontario M.N.R. Midhurst.

Lellinger, David B. 1985. **A Field Manual of the Ferns and Fern Allies of the United States and Canada.** Smithsonian Institution Press, Washington, D.C.

Oldham, M.J. 1996. **Rare Vascular Plants** Ontario Natural Heritage Information Centre, Peterborough, ON.

OMAFRA, 1996 **Bruce County at a Glance, Local Statistics and Facts.** Walkerton Field Office.

OMAFRA, 1996. **Grey County at a Glance, Local Statistics and Facts.** Markdale Field Office.

1977. **Ontario Geological Map: Southern Sheet.** No. 2392. Ontario M.N.R.

Tovell, Walter M. 1965. **The Niagara Escarpment.** Royal Ontario Museum Series. Univ. of Toronto Press.

Tovell, Walter M. 1992. **Guide to the Geology of the Niagara Escarpment.** Niagara Escarpment Commission.

Wagner H., Wagner, F. & Reznicek, A. 1992. **x** *Dryostichum singulare* **(Dryopteridaceae), A new fern nothogenus from Ontario.** Can. J. Bot. Vol. 7.

INDEX OF FERN NAMES

If anyone discovers in a township unit
a fern that is not recorded here,
please inform the M.N.R.
(See form at the back).

COUNTIES OF GREY AND BRUCE
REPORT OF NEW OR RARE PLANT SPECIES

Name of plant: Latin: _____

English: _____

Name of observer: _____

Address: _____

_____ e-mail _____

Telephone number: _____ Fax Number _____

Name(s) of other observers: _____

Date of observation: _____

Number of plants observed? _____

Do you have a photograph? _____ Do you have a pressed specimen? _____

If neither of the above, how was the plant identified?

Location of species: _____

County: _____ Township: _____

Lot and concession numbers: _____

Who owns the property? _____

Precise directions to location (include a sketch map, if possible):

If available give: 1:50 000 Topographic map number: _____

UTM 6 digit grid # & square: _____

Latitude: _____ Longitude: _____

Would you be able to take someone to view the location? _____

Any other information?